AF157942

tredition®

tredition was established in 2006 by Sandra Latusseck and Soenke Schulz. Based in Hamburg, Germany, tredition offers publishing solutions to authors and publishing houses, combined with worldwide distribution of printed and digital book content. tredition is uniquely positioned to enable authors and publishing houses to create books on their own terms and without conventional manufacturing risks.

For more information please visit: www.tredition.com

TREDITION CLASSICS

This book is part of the TREDITION CLASSICS series. The creators of this series are united by passion for literature and driven by the intention of making all public domain books available in printed format again - worldwide. Most TREDITION CLASSICS titles have been out of print and off the bookstore shelves for decades. At tredition we believe that a great book never goes out of style and that its value is eternal. Several mostly non-profit literature projects provide content to tredition. To support their good work, tredition donates a portion of the proceeds from each sold copy. As a reader of a TREDITION CLASSICS book, you support our mission to save many of the amazing works of world literature from oblivion. See all available books at www.tredition.com.

 ## Project Gutenberg

The content for this book has been graciously provided by Project Gutenberg. Project Gutenberg is a non-profit organization founded by Michael Hart in 1971 at the University of Illinois. The mission of Project Gutenberg is simple: To encourage the creation and distribution of eBooks. Project Gutenberg is the first and largest collection of public domain eBooks.

Creatures of the Night A Book of Wild Life in Western Britain

Alfred Wellesley Rees

Imprint

This book is part of TREDITION CLASSICS

Author: Alfred Wellesley Rees
Cover design: Buchgut, Berlin – Germany

Publisher: tredition GmbH, Hamburg - Germany
ISBN: 978-3-8472-2341-2

www.tredition.com
www.tredition.de

Copyright:
The content of this book is sourced from the public domain.

The intention of the TREDITION CLASSICS series is to make world literature in the public domain available in printed format. Literary enthusiasts and organizations, such as Project Gutenberg, worldwide have scanned and digitally edited the original texts. tredition has subsequently formatted and redesigned the content into a modern reading layout. Therefore, we cannot guarantee the exact reproduction of the original format of a particular historic edition. Please also note that no modifications have been made to the spelling, therefore it may differ from the orthography used today.

CREATURES OF THE NIGHT

By the same Author.

IANTO THE FISHERMAN

AND OTHER SKETCHES OF COUNTRY LIFE.

Illustrated with Photogravures. Large Crown 8vo.

The Times.—"The quality which perhaps most gives its individuality to the book is distinctive of Celtic genius.... The characters ... are touched with a reality that implies genuine literary skill."

The Standard.—"Mr Rees has taken a place which is all his own in the great succession of writers who have made Nature their theme."

The Guardian.—"We can remember nothing in recent books on natural history which can compare with the first part of this book ... surprising insight into the life of field, and moor, and river."

The Outlook.—"This book—we speak in deliberate superlative—is the best essay in what may be called natural history biography that we have ever read."

LONDON
JOHN MURRAY, ALBEMARLE STREET

 "THE BROAD RIVER, IN WHICH SHE HAD SPENT HER EARLY LIFE."

CREATURES OF THE NIGHT

A BOOK OF WILD LIFE IN WESTERN BRITAIN

BY ALFRED W. REES

AUTHOR OF
"IANTO THE FISHERMAN"

WITH ILLUSTRATIONS

LONDON
JOHN MURRAY, ALBEMARLE STREET
1905

TO
MYFANWY AND MORGAN
"All life is seed, dropped in Time's yawning furrow,
Which, with slow sprout and shoot,
In the revolving world's unfathomed morrow,
Will blossom and bear fruit."
Mathilde Blind.

PREFACE.

The Editors of *The Standard* have kindly permitted me to republish the contents of this book, and I tender them my thanks.

The original form of these Studies of animal life has been extensively altered, and, in some instances, the titles have been changed.

I am again greatly indebted to my brother, R. Wilkins Rees. His wide and accurate knowledge has been constantly at my disposal, and in the preparation of these Studies he has given me much indispensable advice and assistance.

Similarity in the habits of some of the animals described has made a slight similarity of treatment unavoidable in certain chapters.

I may also remark that, in unfrequented districts where beasts and birds of prey are not destroyed by gamekeepers, the hare is as much a creature of the night as is the badger or the fox.

ALFRED W. REES.

Transcriber's Note: Obvious typographical errors have been corrected, and standardized the hyphenations, otherwise the text has been left as the original

THE OTTER.I.

THE HOLT AMONG THE ALDERS.

I first saw Lutra, the otter-cub, while I was fishing late one summer night. Slow-moving clouds, breaking into fantastic shapes and spreading out great, threatening arms into the dark, ascended from the horizon and sailed northward under the moon and stars. Ever and anon, low down in the sky, Venus, like a clear-cut diamond suspended from one of its many twinkling points, glittered between the fringes of the clouds, or the white moon diffused soft light among the wreathing vapours that twisted and rolled athwart the heavens. In the shelter of the pines on the margin [Pg 4] of the river, a ringdove, awakened by a bickering mate, fluttered from bough to bough; and his angry, muffled coo of defiance marred the stillness of the night. The gurgling call of a moorhen, mingling with the ripple of the stream over the ford, came from the reeds at a distant bend of the river. Nearer, the river, with varying cadence, rose and fell in uneven current over a rocky shelf, and then came on to murmur around me while I waded towards the edge of a deep, forbidding pool. In the smooth back-wash beyond the black cup of the pool a mass of gathered foam gleamed weirdly in the dark; and, further away, broad tangles of river-weed, dotted with the pale petals of countless flowers, floated on the shallow trout-reach extending from the village gardens to the cornfields below the old, grey church.

In one of the terraced gardens behind me a cottager was burning garden refuse; tongues of flame leaped up amid billows of smoke, and from the crackling heap a myriad sparks shot out on every side. While the cottager moved about by the fire, his shadow lengthened across the river, which, [Pg 5] reflecting the lurid glare, became strangely suggestive of unfathomable depths. The moorhen called again from the reeds near the ford, then flew away over the fire-flushed river and disappeared into the gloom; and a water-vole dropped with a gentle plash into the pool.

Casting a white moth quietly over the stream, I noticed beyond the shadows a round mass rising from the centre of the current, moving against the flood, and sinking noiselessly out of sight. There could be no doubt that the shape and motion were those of an otter. To continue my sport would have been in vain with such a master-fisher in the pool, so I reeled in my line, and stood still among the ripples as they circled, muttering, around my knees. Presently the dim form of the otter reappeared a little further up-stream, and I caught sight of a glistening trout in the creature's mouth.

The otter swam, with head just above water, towards the alders skirting the opposite bank, and then, turning sharply, was lost to sight near the overhanging roots of a sycamore. Immediately afterwards, a strange, [Pg 6] flute-like whistle—as if some animal, having ascended from the depths of the river, had blown water through its nostrils in a violent effort to breathe—came from the whirlpool in the dense shadows of the pines: the otter's mate was hunting in the quiet water beyond the shelf of rock. Then a slight, rattling sound on the pebbly beach of a little bay near the sycamore indicated that the animal had landed and was probably devouring the captured fish. The leaping flames of the cottager's fire had been succeeded by a fitful glow, but the moon glided from behind the clouds and revealed a distinct picture of the parent otter standing on the shingle, in company with Lutra, her little cub.

A deserted mansion—to whose history, like the aged ivy to its crumbling walls, clung many a fateful legend—nestled under the precipitous woods in the valley. Time, taking advantage of neglect, had made a wilderness of the gardens, the lawns, and the orchards, which, less than a century ago, surrounded with quiet beauty this [Pg 7] home of a typical old country squire. A few garden flowers still lingered near the porch; but the once well tended borders were overgrown with grass, or occupied with wild blossoms brought from the fields by the hundred agents employed by Nature to scatter seed. Owls inhabited the outhouses, and bats the chinks beneath the eaves. A fox had his "earth" in the shrubbery beyond the moss-grown pathway leading from the door to the gate at the end of the drive. A timid wood-pigeon often flew across from the pines and walked about the steps before the long-closed door. Near the warped window of the dismantled gun-room the end of a large water-pipe formed a convenient burrow for some of the rabbits that played at dusk near the margin of the shrubbery. This water-pipe led to the river's brink; and there, having been broken by landslips resulting from the ingress of the stream during flood, one of the severed parts of the tube formed, beneath the surface of the water, an outlet to a natural chamber high and dry in the bank. The upper portion of the pipe was choked with earth [Pg 8] and leaves washed down from the fields by the winter rains.

In this hollow "oven," on a heap of hay, moss, and leaves, brought hither by the parent otters through an opening they had tunnelled into the meadow, Lutra was born. Her nursery was shared by two other cubs. Blind, helpless, murmuring little balls of fur, they were tended lovingly by the dam.

Soon the thin membrane between their eyelids dried and parted, and they awoke to a keen interest in their surroundings. Their chamber was dimly lit by the hole above; and the cubs, directly they were able to crawl, feebly climbed to a recess behind the shaft, where they blinked at the clouds that sailed beneath the dome of June, and at the stars that peeped out when night drew on, or watched the limpid water as, flowing past the end of the pipe below, it bore along a twirling leaf or rolled a pebble down the riverbed. Occasionally a salmon-pink wandered across from the shallows; for a moment or two the play of its tiny fins was seen at the edge of the pipe; and the cubs, excited by a sight of their future prey, stretched their necks and knowingly held [Pg 9] their heads askew, so that no movement of the fish might escape their observation.

Among flesh-eating mammals of many kinds, the females display signs of intelligence earlier than the males. Lutra being the only female among the cubs, she naturally grew to be the most keenly observant, and often identified the finny visitor before her brothers ventured to decide that it was not a moving twig.

The dam spent most of the day asleep in the "holt," and most of the night fishing in the pools. Inheriting the disposition of their kind, the cubs also were more particularly lively by night than by day. Directly the cold dew-mist wreathed the grass at the entrance of the burrow, they commenced to sport and play, tumbling over each other, grunting and fighting in mimic anger, or pretending to startle their mother directly she entered the pipe on returning at intervals from fishing.

One night, while the cubs were rougher than ever in their fun, Lutra slipped off the platform and fell headlong down the pipe into the stream. But almost before she had time to be frightened she discovered [Pg 10] that to swim was as easy as to play; and she rose to the surface with a faint, flute-like call. She splashed somewhat wildly, for her stroke was not yet perfected by practice. Hearing the commotion and instantly recognising its meaning, the dam dived quietly and swiftly right beneath the cub, and bore her gently back to the platform, where the rest of the family, having missed their companion, had for the moment ceased to romp and fight.

A few nights after this incident, the mother commenced in earnest to educate her young. Tenderly taking each in turn, she carried the nurslings into the water, and taught them, by a method and in language known only to themselves, how to dive and swim with the least possible exertion and disturbance.

Henceforward, throughout the summer, and till the foliage on the trees near the pool, chilled by the rapid fall of the temperature every evening, became thinner in the breath of the early autumn wind, the otter-cubs fished, and frolicked, and slept, or were suckled by their dam. Sometimes the whole family, together with the [Pg 11] old dog-otter, adjourned to the middle of the meadow, and in the tall, dew-drenched grass skipped like kittens, though with comical clumsiness rather than with the agility they displayed in the water. Like kittens, too, the cubs played with their mother, in spite of wholesome chastisement when they nipped her muzzle rather more severely than even long-suffering patience could allow. The dam was at all times loath to correct her offspring, but the sire rarely endured the familiarity of the cubs for long. Directly they became unduly presumptuous he lumbered off to the river, as if considered it much more becoming to fish than to join in the sport of his progeny. Perhaps, indeed, he deemed a change of surroundings essential that he might forget the liberties taken with him by his disrespectful youngsters.

When about three months old, Lutra began to show promise of that grace of form and motion which in later life was to be one of her chief distinctions. Her body, tail, and head gradually lengthened; and, as her movements in the water became more sinuous and easy, she tired less rapidly when fishing. [Pg 12]

Autumn passed on towards winter, the nights were long, the great harvest of the leaves fell thickly on the meadow and the stream, the mountain springs were loosed in muddy torrents, and the river roared, swollen and turbid, past the "holt" under the trailing alder-twigs. The moorhens came back from the ponds where they had nested in April and May; the wild duck and the teal flew south from oversea, and in the night descended circling to the pool; a dabchick from the wild gorge down-river took up his abode in the sedges.

The quick jerk of the dabchick's oar-like wings caused much wonder to Lutra, when, walking on the river-bed, she looked up towards the moonlit sky, and saw the little grebe dive like a dark phantom into the deep hole beneath the rocky ledges of Penpwll. Once the otter-cub, acting under an irresistible impulse, swam towards the bird and tried to seize him. She managed to grip one of his feet, as they trailed behind him while he dived, but the grebe escaped, leaving in the assailant's mouth only a morsel of flesh torn from a claw. [Pg 13]

In the warm evenings of late summer and the first weeks of autumn, the angler usually visited the shingle opposite the water-pipe, and waded up-stream casting for trout. The otter-cubs, grown wiser than when the angler saw them near the sycamore, discreetly stayed at home, for they had been taught to regard this strange being, Man, known by his peculiar footfall and upright walk, as a dreaded enemy scarcely less formidable than the hounds and the terriers that at intervals accompanied him for the express purpose of hunting such river-folk as otters and rats.

As yet Lutra had never seen the hounds, nor, till the following summer, was she to know the import of her instinctive timidity. Roaming, hungry, and venturesome, she had chanced at nightfall to catch a glimpse, during an occasional gleam of moonlight, of a large trout struggling frantically on the surface of the water not far from the angler, had heard the click of the reel and the swish of the landing net, and had concluded that these mysterious proceedings gave cause for fear. [Pg 14]

The end of October drew nigh; and, when the last golden leaves began to fall from the beeches, the angler ceased to frequent the riverside. Henceforward, except when a sportsman passed with his gun, the otters' haunt remained in peace.

Always at break of day, however, when the pigeons left their roosting places in the pines, an old, decrepit woman tottered down the steps from the cottage door to the rock at the brim of the pool, and filled her pails with water. But the creatures felt little alarm: they had become accustomed to her presence in the dawn. Lonely and childless and poor, she knew more than any one else of the otters; but she kept their whereabouts a secret, for the creatures lent an interest to her cheerless, forsaken life, and recalled to her halting memory the long past days when her husband told her tales of hunting and fishing as she sat, a young and pretty girl, at her spinning wheel in the light of the flickering "tallow-dip."

Warm, cloudy weather continued from the late autumn through the winter—except for a few days of frost and snow in [Pg 15] December—so that food was never scarce, and Lutra thrived and grew. The great migration of salmon took place, but she was not sufficiently big and strong to grip and hold these monster fish. Her own weight hardly exceeded that of the smallest of them, so she had to be content with a mixed diet of salmon-fry and trout, varied with an occasional slug or snail that she chanced to find in the meadow. For a brief period after the fall of snow in December, the frost fettered the fields, and the moon shone nightly on a white waste through which the river flowed, like a black, uneven line, between its hoar-fringed banks. Then Lutra, bold in the unbroken stillness of Nature's perfect sleep, climbed the steps leading to a village garden, and searched the refuse heap for scraps discarded from the cottager's meagre board. She even wandered further, crossed the road, and passed under a gate into the fields near the outlying stables of the inn. Here some birds had roosted in the hazels by the fence, and the cub stood watching them, like the fox beneath the desired but distant grapes. [Pg 16]

A rough, mongrel sheep-dog, having missed his master, who had been carousing in the inn that evening, chanced to be trotting homeward to the farm on the hill, and, sniffing at the gate, discovered the cub in the hedgerow. With a mad yell the dog tore through the briars at the side of the gate-post; but Lutra was equally quick, and by the time her enemy was in the field she had dodged under the bars and was shuffling away, as quickly as her short legs permitted, down the garden to the river. The dog turned, crashed back through the briars, and gained rapidly on the otter. He reached her just as she gained the top of the wall that, on a level with the garden, formed a barrier against the river-floods. Lutra felt a sharp nip on her flank, and was bowled over by the impetuous rush of her foe; but she regained her feet in an instant, and jumped without hesitation into the water. The river was shallow where she fell; the dog followed her; and for a moment she was in deadly peril. But before the sheep-dog recovered from his sudden plunge, Lutra swam into the deep water and dived straight for [Pg 17] home, leaving the plucky mongrel standing in the ripples, with a look of almost human disgust and astonishment on his intelligent face. He may have reasoned thus: "Surely I caught that otter. But stay, I must have been dreaming. 'Tis queer, though: I'm in the river instead of on the road to the farm." This, for Lutra, was perhaps the only noteworthy episode of her early life.

The otter-cub was about nine months old when spring came to the valley. The water-weed grew in long filaments from the gravelly shallows. The angler, who had ceased to frequent the riverside at the approach of winter, returned to the pool, but only by day, and then Lutra dozed in her retreat. In the pines on the margin of the river the blue ringdoves were busy constructing the rude makeshift that was to serve the purpose of a nest. Instead of seeking how to construct a perfect dwelling place, these slipshod builders spent most of their hours in courtship. Sometimes, owing to the carelessness of the lackadaisical doves, a dry stick released by bill or claw would fall pattering among the branches, and drop, [Pg 18] with a plash, into the river, where it would be borne by the current past the otter's lair. From every bush and brake along the sparkling stream the carols of joyous birds floated on the morning mists. The first green leaves of the bean peeped in the gardens; the first broods of the year's ducklings launched forth, like heartstrong adventurers, into the shallows by the cottage walls. In the sunny glades the big, fleshy buds of the chestnut and the light-green, tapering sprouts of the sycamore expanded under the influence of increasing warmth. Finches and sparrows, on the lookout for flies, hovered above the ankle-deep drifts of leaf-mould in the lane below the trees, or crossed and re-crossed between the budding boughs. Only a few of these many signs were observed by Lutra, it is true, for she spent the day in hiding. But at dusk she heard the bleating of the lambs, and the musical note of a bell that had been slung round the neck of the patriarch of the flock in order to deter foxes from meddling with the new-born weaklings then under the big ram's care. She was made aware of the presence of spring by the "scent in the [Pg 19] shadow and sound in the light." The hatching of countless flies in the leaf-mould was not watched by the birds only: Lutra also knew that the swarms had arrived; and spring was welcome if only for this.

For months she had fed on lean and tasteless trout exhausted by spawning. Now, instead of lying under stones or haunting the deep basin of the pool, the trout rose to the surface and wandered abroad into the shallows. There the languid fish became fit for food again, and more capable of eluding the occasional long, stern chases of the otter. But Lutra was never disconcerted by the fact that the fish were strong and active; as with all carnivorous creatures, her sporting instincts were so highly developed that she revelled in overcoming difficulties, especially because she felt her own strength growing from day to day. During winter the trout had fed on worms and "sundries." Now, their best and heartiest meals were of flies. Daily, at noon, swarms of ephemerals played over the water, and the trout rose from the river-bed to feed. At first they "sported" ravenously, rising quick and sure to any insect their marvellous vision might [Pg 20] discern. Afterwards they fed daintily, disabling and drowning with a flip of the tail many an insect that fluttered at the surface, and choosing from their various victims some unusually tasty morsel, such as a female "February red" about to lay her eggs. At this time, also, the plump, cream-coloured larvæ of the stone-fly in the shallows were growing within their well cemented caddis-cases and preparing for maturity. So the trout fattened on caddis-grubs and flies, and the otter-cub, in corresponding measure, became sleek, well-grown, and spirited.

In the winter Lutra had imperceptibly acquired the habit of swimming and diving across-stream, just as an old fox, when hunting in the woods, quarters his ground systematically across-wind, and so detects the slightest scent that may be wafted on the breeze. Nature had been specially kind to her; she was fashioned perfectly, and in the river reigned supreme. Her body was long, supple, and tapering; her brown fur was close and short, so that the water never penetrated to her skin and her movements were not retarded as they would have been [Pg 21] had she possessed the loose, draggling coat of an otter-hound. She seemed to glide with extraordinary facility even against a rapid current. Her skin was so tough that on one occasion when, by accident, she was carried down a raging rapid and thrown against a jagged rock, a slight bruise was the only result. Her legs were short and powerful, her toes webbed, and her tail served the purpose of a rudder. Nostrils, eyes, and ears—all were small and water-tight, and set so high on the skull that, when she rose to breathe, little more than a speck could be seen on the surface, unless she felt it safe to raise her head and body further for the sake of ease in plunging deep.

When Lutra was nine months old she caught her first salmon; and, though the fish was only a small "kelt," returning, weak from spawning, to the sea, the capture was a fair test of the cub's prowess and daring. It happened thus. She was walking up the river-bed one boisterous night, when she saw a dark form hovering close to the surface in the middle of a deep pool. Her eyes, peculiarly fitted for watching objects immediately above, quickly detected [Pg 22] the almost motionless fish. The eyes of the salmon were also formed for looking upwards, and so Lutra remained unnoticed by her prey. She stole around the hovering fish, that the bubbles caused by her breathing might make no noticeable disturbance as they rose to the surface, and then, having judged to a nicety the strength of the stream, paddled with almost imperceptible motion towards the salmon. Before the fish had time to flee it was caught in Lutra's vice-like jaws and borne, struggling desperately and threshing the water into foam, to the bank. There the otter-cub killed her victim by severing the vertebræ immediately behind its gills.

Otters well nigh invariably destroy large-sized fish by attacking them in this particular part. And, according to a similar method, stoats and polecats, whenever possible, seize their victims near the base of the brain. In yet another way Lutra proved her relationship to the weasel tribe: just as our miniature land-otters eat only small portions of the rabbits they kill, so the cub was content with a juicy morsel behind the salmon's head—a morsel known among sportsmen as "the otter's bite." [Pg 23]

Soon after the cub had killed her first salmon she separated from her parents and brothers, travelled far down-river, and wandered alone. In the human character, development becomes especially marked directly independence of action is assumed; henceforward parental guidance counts for comparatively little. And so it was with Lutra.

II.

[Pg 24]

Top

THE POOL BENEATH THE FARMSTEAD.

Last year, in autumn mornings, when the big round clouds sailing swiftly overhead reminded me of springtide days and joyous skylarks in the heavens, but when all parent birds were silent, knowing how dark winter soon would chill the world, a thrush, that not long since had been a fledgling in his nest amid a shrubbery of box, came to the fruit-tree near my window, and, in such low tones that only I could hear them, warbled that all in earth and sky was beautiful.

To Lutra, lonely like the thrush, and, like the thrush, not yet aware of pain and hunger, the world seemed bright and filled with happiness. At first, like a young fox that, till he learns the fear of dogs and men, steals chickens from a coop near [Pg 25] which an old, experienced fox would never venture, she was, perhaps, a little too indifferent to danger. In her perfect health and irresponsible freedom, she paid but slight attention to the alarm signals of other creatures of the night.

Up-river, at a bend below a hillside farmstead some distance from our village, is a broad, deep salmon-pool, fringed with alders and willows. Right across the upper end of this pool stretches a broken ledge of rock, over which, in flood, the waters boom and crash into a seething basin whence thin lines of vapour—blue and grey when the day is dull, or gleaming with the colours of the rainbow when the sun, unclouded, shines aslant the fall—ceaselessly arise, and quiver on the waves of air that catch their movement from the restless swirls beneath. But in dry summer weather the ledge is covered with green, slippery weed, the curving fall is smooth as glass, and the rapid loses half its flood-time strength.

This pool, though containing some of the finest salmon "hovers" in the river, is nowadays but seldom fished. Since the [Pg 26] old generation of village fishermen has passed away it seems to have gradually lost its popularity. The right bank of the river above and below the pool is for miles so thickly wooded that anglers prefer to pass up-country before unpacking their rods. From the left bank it is useless for any angler who has not made a study of the pool to attempt to reach the "hovers." Under far more favourable conditions than these, the throw necessary to place a fly on even the nearest of the "hovers" would be almost the longest that could with accuracy be made. But the angler is baffled at the outset by the presence of a steep slope behind him.

I well remember two instances when I was tricked by the self-conceit which led me to suppose that my skill in casting was of no mean order. Once, while the river was bank-high after flood, I happened to be throwing an unusually long line, with careless ease, over the lower end of a pool, where, before, I had never seen a fish. I was, no doubt, thinking of something quite unconnected with fishing, otherwise I should not have wandered thus far from the spot [Pg 27] where I generally reeled in my line. A salmon effectually aroused me by a terrific rush at my fly. I "struck" hard, and the fly, after a momentary check, flew up into the air. I am not one of those anglers who give rest to a salmon in the belief that, after rising, he requires time to recover from his disappointment at having failed to catch the lure. I believe in "sticking to" a fish, perhaps because the first I ever hooked was one I had bullied ceaselessly during the whole of a spring evening. And so I tried hard and often to tempt that sportive fish again; but after the careless, easy casting which resulted in the rise, I could not by any means throw satisfactorily over the tail of the pool. However I tried to do so, the line would double awkwardly as it reached the water, or would curl back into the rapid on the near side of the "hover," or the fly would splash in a most provoking manner as it alighted on the stream. So at last I left the riverside.

Henceforth, I attempted the same long cast whenever I passed the pool. I lost many flies, and never again rose a fish. But I was convinced that I had discovered a "hover" new to the village fishermen, [Pg 28] till my old friend Ianto chaffed me into the belief that the salmon I had seen was a "passenger," and, probably, a "spent kelt" in such a weak condition that for it to stay in the rough water higher up the pool was impossible.

On another occasion, in early days when my ignorance of the river and of fishing sorely troubled both Ianto and myself, as I was wading down-stream along the edge of a pool a grilse rose, "head and tail," about twenty yards below my fly. Using my long gaff-handle as a staff, I walked slowly towards the fish, casting carefully all the way. I was so absorbed in my work that I did not know I was moving into deep water till I found that my wading stockings had filled. I then stopped, and, lengthening my line at each successive "throw," sent my fly nearer and still nearer to the grilse.

How I managed the long, straight cast that presently resulted in my fly passing down the "hover," I do not know. The grilse rose sharply at the lure, but I "struck" too late. I reeled in my line, and after a few minutes began once more to cast. Now, [Pg 29] however, try as I might, I could not get the line out to the distance required; it would not fall straight and true. In desperation I endeavoured to overcome the difficulty by sheer strength. I swung my arms aloft; my old hickory rod creaked and groaned with the increasing strain, then snapped immediately the tension was released with the return of the line; and, a second afterwards, the grilse took my fly and bolted away down-stream.

All caution left me; I was "into a fish"—that was enough. In haste to catch my rod-top as it slipped down the line from the butt, I made one step forward, and fell over head and ears into a deep hole beneath the shelf of rock on which I had been standing. When I recognised what had happened I was clinging to an alder-root near the bank; thence, breathless, I lifted myself till I was safe on a tree-trunk above the pool. My rod and cap were drifting rapidly away; but, after divesting myself of half my dripping garments, I recovered the rod in a backwater below the neighbouring wood. All my line had been taken out, the gut collar had been snapped, and the [Pg 30] fly had undoubtedly been carried off by the grilse.

In those old days of which I have elsewhere written, [1] Ianto and I often resorted to the wide, deep pool under the farm. Sometimes, during summer, we were there before daybreak, fishing for the salmon that only then or in the dusk would deign to inspect our "Dandy" fly. And there, in the summer nights, we frequently captured, with the natural minnow, the big trout that wandered from the rapids to feed in the quiet waters by the alders. Ianto knew the pool so well that even in the darkest night he would wade along the slippery, weed-grown shelf near the raging fall, to troll in the shadows above him. Had the old man taken one false step he would have entered on a struggle for life compared with which my own adventure after hooking the grilse would have been insignificant.

For several months free, happy Lutra made her daytime abode in a "holt" among the alder-roots fringing this pool. She loved in the long winter nights to [Pg 31] hear the winnow-winnow of powerful wings as the wild ducks circled down towards the pool, the whir of the grey lag-geese far in the mysterious sky, and the whistle of the teal and the gurgle of the moorhens among the weeds close by the river's brim.

Crouched on a grassy mound beside the rapids, she could see each movement on the surface of the pool. The wild ducks splattered and quacked as they paddled busily hither and thither, visiting each little bay and reed-clump at the water's edge. Sometimes, surrendering themselves wholly to sport and play, they formed little groups of two or three; and now one group, and then another, would race, half-swimming, half-flying, from bank to bank or from the rock to the salmon "hover" at the lower end of the pool. The otter remembered her experience with the dabchick, and believed that to capture a full-grown duck would tax her utmost strength and cause a general alarm. Once, however, excited by the wild ducks' sport, she slipped quietly from the mound, dived deep, and from the river-bed shot up in the midst of the birds just as they had congregated [Pg 32] to settle a point of difference in a recent event, and to discuss a second part of their sports' programme for the night.

As the birds, panic-stricken, scattered on every side, and, following each other in two long lines that joined in the form of a wedge, flew up into the starlit sky, Lutra watched them eagerly for a few moments; then, without a ripple, she sank below the surface and returned to her watch on the mound. For a while after the ducks had left the pool, nothing could be heard but the ceaseless noise of falling water. But as the night drew on, a moorhen ventured from the shelter of the alders, and, like a tiny, buoyant boat, launched out into the pool. The otter, with appetite whetted by recent sport among the ducks, again left her hiding place and silently vanished into the stream. Borne by the current, she reached, with scarcely an effort, a point in the swirling depths from which she could catch a glimpse of the dim outline of the floating bird. Then, rising swiftly, she gripped the moorhen from beneath, dived across to the "hover," and, having killed and skinned her prey, feasted at leisure. [Pg 33]

There were times in the second summer of her existence when Lutra, like the wild ducks, seemed to abandon every thought of the possibility of danger. Simply for the love of exercise and in enjoyment of the tranquil night, she played about the pool till the dawn peeped over the hills; then, tired of her frolic, she sought her secret "holt," and, curling her tail about her face and holding her hind-paws closely between her fore-paws, fell asleep.

While she gambolled in the water, even her quickest movements were as graceful as those of a salmon stemming the rapids and leaping into the shallows above the rock. Diving into the depths, she avoided with scarcely an effort the tangled roots and branches, that, washed thither by the floods, had long been the dread of anglers when heavy fish were hooked. Ceasing all exertion as she turned into the current, she floated to the surface and was borne away down-stream. She swam at highest speed from the tail to the throat of the pool, and drifted idly back to the place from which she had started; then, changing her methods, she skirted [Pg 34] slowly the edge of the current, and with one long, straight dive shot down from the head of the rapids to the still water near her "holt."

From playing thus about the pool, the otter learned the power of the current, and how it hastened or retarded her while she pursued her prey. But most of all, during the hours of the placid night, she delighted to frolic in the torrent immediately below the rock, where, matching her strength against that of the river, she leaped and dived and tumbled through the foam, or, lying on her back amid a shower of spray, stretched wide her limbs and suffered the whirlpool to draw her, unresisting, into its vortex deep beneath the fall.

Lutra sometimes noticed, while she drifted with the current, that the scent of her kindred lay strong at the surface not far from her "holt." One still, moonlit night the scent indicated that several full-grown otters had at intervals come from the trout-reaches downstream, and had landed in a reed-bed at the lower end of the pool. It led away from the river through the valley, along by a number of stagnant [Pg 35] ponds in an old garden near the farm, and thence to a point beyond a bend where the river flowed almost parallel to its course at the pool. As the otter, inquisitively following the line of the scent, came to the ponds, she heard the croaking of countless frogs hidden in the duckweed that lay over the entire surface of the water. Lutra made ample use of the opportunity for a feast—frogs were the greatest delicacies known to her, and she had never before found them to be so plentiful. Dawn was breaking when, in her onward journey, she reached the river; so she drifted around the bend, dived over the fall, and returned to her home beneath the alder-roots.

It happened that the otters whose "spur" (footprints) Lutra had followed to the frog-ponds retraced their steps towards the pool, and in doing so suddenly discovered that the scent of a man lay strong on the trodden grass. A villager, knowing the eagerness with which otters seek for frogs, and that they often cross a narrow neck of land at the bend of a stream, had for a time kept watch at the lower end of the old farm garden. He was anxious that the [Pg 36] hounds, which, on the previous day, had arrived at the village, should enjoy good sport during their stay in the neighbourhood. But he saw nothing of the animals he had come to watch; as soon as they detected his whereabouts they retreated hastily to the pond at the upper end of the garden, gained the river, and, like Lutra, swam homewards around the bend. But, less familiar than Lutra with the strength of the current, they left the water as they approached the fall, and crept through the deep shadows of the alder-roots till they reached a point at some distance beyond the pool.

These events of the night were of the utmost importance to the otters as connected with the events of the morrow. During the early morning the villager paid a second visit to the garden, and examined closely the soft mud at the margin of the ponds. The remains of the otters' feast—the skins and the eyes of frogs—lay in several places, and, near the largest of the ponds, the otters' "spur" showed clearly that the animals had for some time been busy there. Taking a [Pg 37] straight course to the river above the pools, the watcher again detected the marks of the otters on the sloping bank. By the riverside below the garden, however, he failed to observe any further sign, and so concluded that the animals had probably left the water at the opposite bank.

When, later, the Hunt crossed the bridge on its way up-stream, the villager told his story to the Master, who immediately led his hounds over the hill-top in the direction of the ponds. This unexpected movement drew the followers of the Hunt away from the river; they imagined that the hounds were to be taken across country to a well known gorge where, during a previous season, good sport had been obtained.

At the farm, the Master, leaving the hounds to the care of the whippers-in, waited till the villagers and the farmers had congregated in the yard. He then addressed the crowd, telling them that otters had visited the garden during the night and probably were still in hiding there, and that, if good sport were desired, it would be wise for his followers to form two groups and watch the fords above and below [Pg 38] the river-bend, while he, alone, accompanied the hounds to the garden; his chief reason, he said, for pointing out to them the advisability of leaving him was that if an otter still remained near the pond it should be given every chance of reaching the river without molestation. The crowd, recognising the wisdom of the Master's remarks, moved off with the whippers-in to the fords; and, when all was in readiness, the pack was led into the garden. One, and another, and yet another of the "young entry" soon gave tongue; then, after a minute's deliberation, an old, experienced hound raised his head from the rushes, uttered a single deep, clear note, climbed the garden hedge, and galloped across the meadow towards the river.

The rest of the hounds speedily found the line of the "drag," but all came to a check at the water's edge. They were taken back to the ponds, and thence to the pool by the farm, but the scent was weak above the waterfall. They again "cast" to the upper end of the garden, and onward to the river. Carefully searching every hole and corner in the bank, they drew down-stream [Pg 39] around the bend, and at last struck the scent of the otters among the reeds below the pool. Lutra heard them tearing madly past, heard also the dull thud of human footsteps above her "holt," but she discreetly remained close-hidden in her sleeping chamber. For hours, in a pool beyond the trout-reach, her visitors of the previous night were hustled to and fro, and frequent cries of "Gaze! gaze!" and "Bubble avent!" mingled with the clamour of the hounds. Then the commotion seemed suddenly to subside. After an interval the hounds splashed by once more among the alder-roots, and the thud of human footsteps resounded in the "holt." In the silence that followed, Lutra, reassured, dived from her "holt," and, paddling gently to the surface, saw the last stragglers of the Hunt climbing the slope towards the farm.

That night no otter from the down-stream trout-reach wandered to the salmon-pool beneath the farm. The water-voles and the moorhens were unusually alert as they swam hither and thither in the little bays along the edge of the current. The fear of [Pg 40] man and his loud-tongued hounds rested, like a spell, on the creatures of the river. Even Lutra felt its power; but when the scent of her foes became so faint as to be lost in the fragrance of the meadow-sweet along the river-bank, she ventured into the old garden, and, on returning to the pool, played again in the raging water by the fall.

Top

III.

[Pg 41]

THE GORGE OF ALLTYCAFN.

When Lutra had attained her full size and strength she was wooed and won by a young dog-otter of her own age, and lived with him in a "holt" among the great rocks of Alltycafn. Now, again, the Hunt arrived in the neighbourhood. It was a lovely morning in May. The sun shone brightly; the leaves were breaking from their sheaths; the birds sang blithely in the trees. Suddenly the otters, resting in their "holt," were awakened by a loud commotion— the sounds of hurrying feet, reverberating in the chamber among the boulders, and then the music of the shaggy hounds, varied occasionally by the yap-yap of the terriers. The noise drew [Pg 42] rapidly nearer. Presently a man, in red stockings and vest, blue breeches and coat, and a blue hunting cap bearing an otter's "pad" mounted in silver, poked among the boulders with a steelshod pole. The dog-otter was now thoroughly alarmed. He rushed from his lair, dived straight into the stream, headed through the seething current, and rose in the adjoining pool. Threatened by a hound, he dived again, walked over the gravel, and swam under the gnarled roots of an oak. The members of the Hunt stood watching the bubbles, filled by his breath, as they floated up and broke. The hounds swam pell-mell in hot pursuit, and the otter was forced to turn up-stream. Moving cautiously under the rocky ledges, he regained the "holt," where his terrified mate awaited his return. Sorely pressed, the dog-otter hid close, hoping to baffle his relentless pursuers. But a bristling, snarling terrier soon came down the shaft from the bank. Maddened, and courageous with the fury of despair, Lutra seized the intruder by the muzzle, and, in the combat that ensued, sorely mangled her assailant's lips and nostrils. Then, as her mate dived [Pg 43] out once more and swam down-stream, she also left the chamber. She rose immediately among the surrounding boulders, and hid in the furthest recess. With nostrils, eyes, and ears raised slightly above the surface of the water, she stayed there, unseen and hardly daring to breathe, and, with strained senses watched closely every movement of hounds and hunters.

Fortunately for Lutra, the arch of the boulders below was shaped so peculiarly that the scent of her breath and body was sucked into a cavity and carried down-stream, and, passing beneath the stone, mingled, at the raging cataract near the rock, with air in the bubbles formed by the tumult of the waters. These bubbles, instead of bursting, were drawn into the vortex of a little whirlpool; and the keen-nosed hounds, though suspicious, could form no definite opinion as to the presence of a second otter among the rocks. The terrier knew the secret, but he had been put out of action and sent off, post haste, to the nearest veterinary surgeon. Lutra saw her tormentors—some of them of the pure otter-hound breed, some half otter-hound, [Pg 44] half fox-hound, and others, again, fox-hounds trained for otter-hunting—rushing backwards and forwards in the water and on the bank. Another terrier, led by a boy, strained at his leash near the river's brink. Women, dressed, like the men, in smart scarlet and blue, and as ready to wade into the stream as the huntsman himself, stood leaning on their otter-poles not far away. At the fords above and below the "pool," the dog-otter's egress was barred by outposts of the enemy standing and splashing, in complete lines, from bank to bank. Once, in despair, the otter actually tried to break through the human chain; but a hunter "tailed" him for a moment, and then dropped him into the deeper water beyond the ford.

The sound of horn, the shouts of men, the deep-toned notes of great hounds, the shrill yapping of eager terriers, and the splashing and the plunging on every side, almost bewildered Lutra. Fearing to move from her shelter, she floated in the deep basin of the hidden pool beside the cataract, till at last the commotion gradually subsided, and hounds and hunters passed out of sight down-stream.

Lutra awaited her mate's return, but in [Pg 45] vain. Not till night did she venture from her hiding place. When, however, the stars appeared, she swam wearily from pool to pool, calling, calling, calling. She explored each little bay, each crevice in the rock. She walked up the dry bed of a tributary brook, and searched among the gnarled roots and the dry, brown grass fringing the gravelly watercourse. She skirted the meadows and the rocks where the hunters had beaten down the gorse and the brambles near her home; thence she returned to the pool. Hitherto she had loved the placid night; to her the stillness was significant of peace. But now that stillness was full of sadness, and weariness, and monotony. The shadows were deep within the gorge; from the distant woods the hoot of an owl mocked her loneliness. She heard no glad answering cry. Still calling, calling, calling, she floated through the shadows, and out into the moonlight shimmering on the placid water below the gorge; but she sought and called in vain.

Lutra spent the rest of that year in widowhood. In consequence of her fight with the terrier, and also because of her grief, her two little cubs were still-born. [Pg 46]

Midsummer came, and the shallows were almost choked with weeds. The countryside experienced a phenomenal period of drought, and for weeks the river seemed impure and almost fetid. Night after night, and steadily travelling westward, Lutra took short cuts across country from pool to pool. Late in July she reached the estuary of the river; and for the remaining months of summer fished in the bay, finding there a pleasant change in her surroundings. Once she was chased by some men in a boat, who shot at her as she appeared for an instant to breathe. Quick and watchful, she dived at the flash, and the pellets fell harmlessly overhead. Again she rose, and again she dived just in time to avoid the leaden hail. Then she doubled back towards the estuary, and the baffled sportsmen sailed away across the bay. As autumn came once more she returned to the river, and fed chiefly on the migrating eels that swarmed in the hollows near the bank. Presently, by many a nightly journey, she gained the upper reaches, where she lived, till the following spring, close to her old home.

The winter was long and severe. In [Pg 47] January, the fields were buried in snow, the roads were as smooth and hard as glass, and the well-remembered pool beneath the pines was almost covered with a great sheet of ice. At this time another young dog-otter began to show Lutra considerable attention. The village children often saw the pairing otters, for the animals, hard pressed, had perforce to fish by day instead of by night. All night the trout lay dormant under the stones in the bed of the river, and only at noon did they rise to the surface on the lookout for hardy ephemerals that, in a short half hour of warmth, were hatched at the margin of the stream. Lutra and her companion followed the fish, and afforded a rare, unexpected sight as, bold with hunger, they ascended to breathe between the sheets of ice in the pool by the village gardens. At night the otters wandered over the snow, and sometimes visited the hillside farms. There, among rotting refuse-heaps, they discovered worms and insects sheltering in genial warmth. When exceptionally hungry, Lutra and her mate would dig into the chambers of the mole and the field-vole in the meadows, and search [Pg 48] ravenously for the inmates. Among the roots of the spreading oaks, the otters found, also, such tit-bits as the larvæ of moths and beetles. A starved pigeon fallen from the pine-boughs; an occasional moorhen weak and almost defenceless; a wild duck that Lutra had captured by darting from beneath a root while the indiscreet bird was feeding, head downwards, at the river's brink—these were among the varied items of the hungry otters' food. Life was indeed hard to maintain. And, to crown the misfortunes of the ice-bound winter, Lutra's matrimonial affairs were once more cruelly disturbed: her mate was caught in a steel trap that Ned the blacksmith had baited and laid in the meadows near the village bridge. He had marked the otters' wanderings by their footprints in the snow, and had then matured his plans.

The calamity occurred one morning, just before daybreak, as the otters were returning to the river from a visit to a hen-coop, where they had found an open door and a solitary chicken. The trap was placed on the grass by the verge of the stream. A [Pg 49] light fall of snow had covered it, but had left exposed the entrails of a chicken which, by coincidence, formed the tempting bait. Distressed and perplexed, Lutra stayed by the dog-otter, trying in vain to release him from his sufferings. The trapped creature, beside himself with rage and fear and pain, attempted to gnaw through his crunched and almost severed foot; but as the dawn lightened the east, and before the limb could be freed, Ned the blacksmith was to be seen hurrying to the spot. Lutra dived out of sight, and, unable to interpose, watched, for a second time, a riverside tragedy. Her attachment, however, had not been of so ardent a nature that bereavement left her disconsolate. Before April she forgot her trapped friend, and was mated again.

Lutra's new spouse had his home in the tributary stream of a neighbouring valley. So, when the snows had melted and the rime no longer touched with fairy fingerprints the tracery of the leafless boughs, and when Olwen the White-footed had come once more into the valley called after her name, Lutra forsook the broad river in which she had spent her early life, and, [Pg 50] with her companion and a promising family, lived contented under the frowning Rock of Gwion, secure in peace and solitude, at least for a season, from the shaggy otter-hounds.

[Pg 51]

Top

THE WATER-VOLE.

[Pg 53]

I.

OUR VILLAGE HOUNDS.

Not many years ago the pleasures of life among my neighbours here in the country were simpler and truer than they are to-day. Perhaps in that bygone time money was more easily made, or daily need was met with smaller expenditure. It may be, too, that family cares were then less pressing, or that a prolonged period of general prosperity had been the privilege of rich and poor alike in this green river-valley around my home. In those days, to which I often look back with regretful yearning, everybody seemed to have leisure; the ties of friendship were not severed by malicious gossip; old and young seemed to realise how good it was to have pleasant acquaintanceships and to be in the sunshine and the open air. Fathers played [Pg 54] with their children in the street: one winter morning, when, after a heavy fall of snow and a subsequent frost, the ground was as slippery as glass, I watched a white-haired shopkeeper, lying prone on a home-made toboggan, with his feet sprawling behind for rudder, steer a load of merry youngsters full tilt down a steep lane behind his house. The sight was so exhilarating that I also forgot I was not a child; and on the second journey I joined the sportive party, and came to grief because the shopkeeper kicked too quickly at a turn in the course and sent me with a double somersault into the ditch.

It happened in those days that in the miscellaneous pack of mongrels our village sportsmen gathered together when they went rabbit-shooting among the dense coverts of the hillsides were two exceptionally clever dogs — a big, shaggy, bobtail kind of animal, and a little, smooth-coated beast resembling a black-and-tan terrier.

The big dog, Joker, lived at a farm in the village, and, during the leisure of summer, when rabbiting did not engage his attention, took to wandering by the river, joining the bathers in their sport and poking his nose [Pg 55] inquisitively under the alder-roots along the bank. While, one sultry noon, the fun in the bathing pool was at its height, Joker routed an otter from a hiding place near which the bathers were swimming with the current, and a terrific fight took place in the shallows before the *dwrgu* made good his escape. The dog was found to have been severely worsted in the fray, and was taken home to be nursed till his wounds were healed. Meanwhile, Joker's fame as an otter-hound was firmly established in the village, and he was regarded as a hero.

The little dog, Bob, lived at the inn, and for years his droll ways endeared him to every villager, as well as to every angler who came to "the house" for salmon-fishing. He loved nothing better than a friendship with some unsuspecting fisherman whom he might afterwards use to further his own ends. The sight of a rod placed by the door in the early morning was sufficient promise of a day's continuous enjoyment; the terrier assumed possession of the rod at once, and kept all other curs at a distance. On the appearance of the sportsman, he manifested such unmistakable delight, and [Pg 56] pleaded so hard for permission to follow, that, unless the sportsman happened to be one whose experiences led him to dislike the presence of a fussy dog by the riverside, the flattery rarely failed of its object. Once past the rustic swing-bridge at the lower boundary of the waters belonging to the inn, Bob left the sportsman to his own devices, and stole off into the woods to hunt rabbits. Unfailingly, however, he rejoined his friend at lunch.

On Sundays, knowing that the report of a gun was not likely then to resound among the woods, and depressed by the quietness and disappointed by the nervous manner with which everybody well dressed for church resented his familiarities, he lingered about the street corners—as the unemployed usually do, even in our village— till the delicious smells of Sunday dinners pervaded the street. The savoury odours in no way sharpened his appetite, for at the inn his fare was always of the best; but they indicated that the time was approaching when the watchmaker and the lawyer set out together for their long weekly ramble through the woods. Bob knew what [Pg 57] such a ramble meant for him. The watchmaker's dog, Tip, was Bob's respected sire, and Tip's brother, Charlie, dwelt at a house in "The Square." Bob, scenting the Sunday dinners, went at once to call for Charlie, and in his company adjourned to the lane behind the village gardens, till the watchmaker and the lawyer, with Tip, were ready for their customary walk.

When the water was low and anglers seldom visited the inn, Bob, during the summer week-days, followed Joker's course of action, and attached himself to a bathing party frequenting a pool below the ruined garden on the outskirts of the village. There, like Joker, he searched beneath the alder-roots, but without success as far as an otter was concerned. However, he vastly enjoyed himself digging out the brown rats from their holes along the bank not far from a rick-yard belonging to the inn, and then hunting them about the pool with as much noise and bustle as if he were close at the tail of a rabbit in the furze. He was so fond of the water that he became a rapid, untiring swimmer; and the boys trained him, in intervals of rat-hunting, to dive to the [Pg 58] bottom of the river and pick up a white pebble thrown from the bank. Like Joker, also, he gained a name for pluck and ability; and one night the village sportsmen, at an informal meeting in the "private room" of the inn, decided to hunt in the river on Wednesday evenings, with Bob and Joker at the head of a pack including nearly every game-dog in the near neighbourhood, except certain aristocratic pointers and setters likely to be spoiled by companionship with yelping and excited curs.

A merrier hunting party was never in the world. They would foregather in the meadow below the ruined garden: the landlord, whose home-brewed ale was the best and strongest on the countryside; the curate, whose stern admonitions were the terror of evil-doers; the farmer, whose skill in ferreting was greater than in ploughing; the watchmaker, whose clocks filled the village street with music when, simultaneously, they struck the hour; the draper, whose white pigeons cooed and fluttered on the bridge near his shop; the solicitor, whose law was for a time thrown to the winds; and a small crowd of boys ready to assist, [Pg 59] if required, in "chaining" the fords. There they would "cry" the dogs across the stream till the valley echoed and re-echoed with shouts and laughter.

The first hunt was started in spirited fashion; the men walked along the bank thrusting their sticks into crevices and holes; but only Joker and Bob entered the water, and rats and otters for a while remained discreetly out of view. Near a bend of the stream, however, Bob surprised a rat secreted by a stone, and, forcing it to rush to the river, followed with frantic speed. Here, at last, was a chase; the other dogs all hurried to the spot, and the landlord, swinging his otter-pole, waded out to perform the duties of huntsman with the now uproarious pack. His action proved infectious—watchmaker, draper, lawyer, and curate splashed into the shallows to help in keeping the rat on the move; and fun was fast and furious till the prey, fleeing from a smart attack by Bob, was captured by a spaniel swimming under a big oak-root between the curate and the bank.

I hardly think I have enjoyed any sport so well as those Wednesday evening hunts in the bygone years, when life was [Pg 60] unshadowed and each sportsman of us felt within him the heart of a child. So great was our amusement that the village urchins instituted a rival Hunt in the brooks on Saturdays; they notched their sticks for every "kill," and boasted that they beat us hollow with the number of their trophies.

We had several adventures with otters, but the creatures always, in the end, eluded us, and we soon were of opinion that smaller fry were capable of affording better fun. Some seasons afterwards, when our Hunt was disbanded, the shopkeepers' apprentices continued, with the youngsters, to work our mongrel hounds; but eventually Joker's death from the bite of an adder put an end to their pastime, for the bobtail and the terrier were the only possible leaders of the nondescript pack.

Bob, the terrier, was always the most interesting of our hounds. He manifested a disposition to use the other dogs to serve his purposes, just as he used the unsuspecting fishermen if he wished to go hunting in the woods. When with me after game on the upland farms, he often seemed to forget entirely that I had taken him to hunt, not for his own amusement only, [Pg 61] but also for mine. Directly he discovered a rabbit squatting in a clump of grass or brambles, perhaps ten or a dozen yards from a hedge, he signalled his find by barking so incessantly that my spaniels hastened pell-mell to the spot. This was just as it should be—for Bob. Dancing with excitement, he waited between the clump and the hedge till the spaniels entered and bolted the rabbit; then he tore madly in close pursuit of the fleeing creature, and my chance of a shot was spoiled through the possibility of my hitting him instead of his quarry.

By the riverside, his tricks were precisely similar. Seeing a moorhen dive, he would call the dogs around him, so that they might bring the bird again to the surface and thus afford him sport. The moorhen, meanwhile, invariably escaped; yet Bob failed to understand that he was the only diver in the pack.

His antics were comical in the extreme if a vole eluded him by diving to the lower entrance of its burrow beneath the surface of a backwater. Having missed his opportunity, but unable to comprehend how he had missed it, the terrier left the water, stood on the roots of a tree over the entrance to [Pg 62] the vole's burrow, and furiously barked instructions to his companions swimming in the pool. Disgusted at last by their inattention to his orders, he plunged headlong into the stream and vanished for a few moments; then he reappeared, proud of his superior bravery, sneezing and coughing, and with a mouthful of stones and soil torn from the bank in his desperate efforts to force his way to the spot whither the object of the chase had gone from view.

Bob long survived the big dog Joker, and in his old days loved as well as ever the excitement of a hunt. His originality was preserved to the end; stiffened by rheumatism and almost choked by asthma, he always, when in search of rabbits, ran up-hill and walked downhill, thus losing both energy and breath that might with advantage have been kept in reserve.

With the passing of the years, many changes have occurred to sunder the friendships formed during those boylike expeditions. I smile when I think how impossible it would be, now that the veneer of town life has been thinly spread over the life of our village, for the man of law to go wading, with tucked-up [Pg 63] trousers, after rats; how impossible, also, for him to frequent with me the bathing pool, as was sometimes his wont, and swim idly hither and thither, while the moon peered between the trees and the vague witchery of the summer night filled his spirit and my own. My youthful feelings, long preserved, have been irrevocably lost; and yet, if only for memory's sake, I would willingly hunt with him again, and, when night had fallen, swim with him once more in the dim, mysterious pool below the garden. But the old hunting party could never be complete. Death makes gaps that Time fails to fill.

Those evenings were delightful, not only because of unrestrained mirth and innocent sport, but also because we took a keen interest in our surroundings, seeing the world of small things by the river-bank with eyes such as belonged to anglers and hunters of the old-fashioned, leisurely school. They marked for me the beginning of a pleasant study of the water-voles that lived in their burrows on the brink of the river, and were sometimes hunted as persistently as were the brown rats, but far more frequently eluded our hounds [Pg 64] than did the noxious little brutes we particularly desired to destroy.

Wherever they take up their quarters, about the farmstead during winter or in the open fields during summer, brown rats are an insufferable nuisance. There is no courtesy or kindness in the nature of the rat; no nesting bird is safe from his attacks, unless her home is beyond his reach in some cleft of a rock that he cannot scale or in some fork of a tree that he cannot climb. He is a cannibal—even the young and the sick of his own kind become the victims of his rapacious hunger—and he will eat almost anything, living or dead, from the refuse in a garbage heap to the dainty egg of a willow-wren in the tiny, domed nest amid the briars at the margin of the river.

The water-vole is often called, wrongly, the water-rat, but it is of very different habits, and is well nigh entirely a vegetable feeder, and one of the most charming and most inoffensive creatures in Britain. To the close observer of Nature, differences in the character of animals—even among the members of one species—soon become apparent. I was struck with manifestations of such unlikeness when I kept small communities of ants [Pg 65] in artificial nests between slips of glass, so as to watch their doings in my hours of leisure. One nest of yellow ants contained at first a dozen workers and a queen; and when I began to study them I used to mark with minute spots of white the bodies of the particular ants under observation. These spots would remain till the ants had time for their toilet and either licked themselves clean or were licked clean by sympathetic companions. At the outset I found that under a magnifying glass two of the dozen workers were readily distinguishable from the others because of their size and shape. Gradually, by detecting little peculiarities, I could single out the ants, and so had no need to mark my tiny pets in order to follow their movements, except on occasions when they clustered round the queen, or rested, gossiping in little groups, here and there in the rooms and passages of their dwelling. One ant was greedy, and, if she was the first to find a fresh drop of honey I had placed outside the nest, would feed to repletion without ever thinking of informing her friends of her discovery. At such times she even became [Pg 66] intoxicated, and I fancied that, when she did at last get home, eager enquiries made as to the whereabouts of the nectar met with incoherent replies, since the seekers for information generally failed to profit by what they were told, and had to cast about aimlessly for some time before finding the food. I also observed that another ant was perfectly unselfish, and not only would inform her companions directly she discovered honey, but would assiduously feed the queen before attending to her own requirements. And so my pets were separately known because of faults and failings or good qualities that often seemed quite human.

A certain vole, living in the river-bank near the place where the villagers met to hunt, was not easily mistaken for one of his fellows. Whereas the general colour of a water-vole's coat—except in the variety known as the black vole—is greyish brown, which takes a reddish tinge when the light glances on it between the leaves, his was uniformly of a dark russet. In keeping with this shiny russet coat, his beady black eyes seemed to glisten with unusual lustre; and so it happened that the question, "I wonder [Pg 67] if Brighteye is from home?" was often asked as we sent our hounds to search among the willows on the further bank; and later it became a custom for the Hunt, before the sport of the evening was begun, to pass up-stream for a hundred yards or so in order that he might be left in peace.

He was quite a baby water-vole when first I made his acquaintance, but the colour of his coat did not change with the succeeding months, and, evening after evening, when the noisy hounds were safe at home or strolling about the village street, I would quietly make my way back to his haunt, and, hidden behind a convenient tree, carefully watch him. In this way I learned many secrets of his life, noticed many traits in which he differed from his companions, and could form a fairly accurate idea of the dangers that beset him, and of the joys and the sorrows that fell to his lot during the three years when his presence was familiar as I fished in the calm summer twilight, or lay motionless in the long grass near the place where he was wont to sit, silent and alert, before dropping into the backwater and beginning the work and the play of the night.

II.

THE BURROW IN THE RIVER-BANK.

The first faint shadows of dusk were creeping over the river when Brighteye, awakened by a movement on the part of his mother, stole from his burrow into the tall grass at the edge of the gravel-bank by the pool. His home was situated in a picturesque spot between the river and a woodland path skirting the base of a cliff-like ascent clothed with giant beeches and an under-garment of ferns and whinberry bushes. Alders and willows grew along the gravel-bank, and through the moss-tangles among the roots many a twisting, close-hidden run-way led upwards to what might be called a main thoroughfare, in and out of the grass-fringes and the ivy, above high-water mark. This road, extending from the far-off tidal estuary to the river's source in the wild mountains to the north, communicated with all the dwellings of the riverside people, and had been kept clear for hundreds of years by wandering voles and water-shrews, moorhens, water-rails, and coots, and, in recent days, by those unwelcome invaders, the brown rats. Here and there it merged into the wider trail of the otter. Sometimes, near a hedge, it was joined by the track of rabbits, bank-voles, field-voles, weasels, and stoats, and sometimes, where brooks and rills trickled over the stones on their way to the river, by other main roads that had followed the smaller water-courses from the crests of the hills.

Brighteye's home might be likened to a cottage nestling among trees at the end of an embowered lane well removed from busy traffic; it contained four or five chambers wherein the members of his family dwelt; and to Brighteye the tall reeds and the bramble thickets were as large as shrubs and trees are to human beings. And, like a sequestered cottager, he knew but little about the great road stretching, up-stream and down-stream, away from his haunts; he was content with his particular domain—the pool, the shallows beyond, a hundred yards of intersected lanes, and the wide main road above the pool and the shallows.

For a time Brighteye sat at the edge of the stream, alert for any sign of danger that might threaten his harmless existence. Then playfully he dropped into the pool, dived, sought the water-entrance to his house, climbed inside his sleeping chamber, and thence to the bank, where again he sat intently listening as he sniffed the cool evening air. A quick-eyed heron was standing motionless in a tranquil backwater thirty yards up-stream; the scent of the bird was borne down by the water, and the vole caught it as it passed beneath the bank. But he showed no trace of terror; the heron was not near enough to give him any real cause for alarm. The rabbits stole down through the woods, the undergrowth crackled slightly as they passed, and one old buck "drummed" a danger signal. Instantly the vole dived again, for he interpreted the sound to mean that a weasel was on the prowl; and, as he vanished, the first notes of a blackbird's rattling cry came to his ears. [Pg 71]

Brighteye stayed awhile in his burrow before climbing once more to the upper entrance. Then cautiously he advanced through the passage, and gained his lookout station. Not the slightest taint of a weasel was noticeable on the bank; so, regaining confidence, he sat on his haunches, brushed his long, bristly whiskers with his fore-feet, and licked his russet body clean with his warm, red tongue. Then he dropped once more into the pool, and swam across to a reed-bed on the further margin. There he found several of his neighbours feeding on roots of riverside plants. He, too, was hungry, so he bit off a juicy flag at the spot marking the junction of the tender stalk with the tough, fibrous stem; then, sitting upright, he took it in his fore-paws, and with his incisor teeth—shaped perfectly like an adze for such a purpose—stripped it of its outer covering, beginning at the severed edge, and laying bare the white pith, on which he greedily fed.

While thus engaged, he, as usual, watched and listened. The spot was dangerous for him because of its distance from the stream, and because the water immediately beyond [Pg 72] was so shallow that he could not, by diving, readily escape from determined pursuit.

His meal was often interrupted for a few moments by some trifling incident that caused alarm. A moorhen splattered out from the willow-roots, and Brighteye crouched motionless, till he recognised that the noise made by the clumsy bird was almost as familiar to him as the rustle of the reeds in a breeze. The blue heron rose heavily from the backwater, and winged his slow flight high above the trees. Here, indeed, seemed reason for fear; but the great bird was not in the humour for killing voles, and soon passed out of view. Now a kingfisher, then a dipper, sped like an arrow past the near corner of the pool; and the whiz of swift wings—unheard by all except little creatures living in frequent danger, and listening with beating hearts to sounds unperceived by our drowsy senses dulled by long immunity from fear—caused momentary terror to the water-vole. Each trifling sight and sound contributed to that invaluable stock of experience from which he would gradually learn to distinguish without hesitation between friends and foes, and be freed from the [Pg 73] pain of needless anxiety which, to Nature's weaklings, is at times almost as bitter as death.

Brighteye was fated to meet with an unusual number of adventures, and consequently to know much of the agony of fear. His russet coat was more conspicuous than that of his soberly gowned companions, and he was on several occasions marked for attack when they escaped detection. But he became the wisest, shyest, most watchful vole along the wooded river-reach, and in time his neighbours and offspring were so influenced by his example and training that a strangely furtive kindred, the wildest of the wild, living in secrecy—their presence revealed to loitering anglers only by tell-tale footprints on the wet sand when the torrent dwindled after a flood—seemed to have come to haunt the river bank between the cottage gardens and the swinging bridge above the pool where Brighteye dwelt.

Though Brighteye's distinctive appearance attracted the notice of numerous enemies, his marked individuality was not wholly a misfortune, since it aroused my kindly interest, and thus caused him to be spared by the village hunting party. [Pg 74]

As he sat in the first shadows of evening among the reeds and the rushes, the kingfisher and the dipper, by which a few minutes before he had been startled, flew back from the direction of the village gardens; and he quickly decided, while watching their flight, that somehow it must be connected with the dull, but now plainly audible, thud of approaching footsteps on the meadow-path. The buck "drummed" again, then the rustling "pat, pat" of the rabbits ceased in the wood, and one by one the adult voles feeding in the reed-bed slipped silently into the shallows and disappeared.

Brighteye was loath to relinquish the juicy rush that he held in his fore-paws, but the signs of danger were insistent. After creeping through the reeds to the water's edge, he proceeded a little way down the bank till he came to a spot where the view of the meadow-path was uninterrupted. His sight was not nearly so keen as his scent and hearing were, but he discerned, in a blur of dim fields, and rippling water, and evening light peering through the willow-stoles, a number of unfamiliar moving objects. He heard quick, uneven footsteps, [Pg 75] and, now and then, a voice; and was aware of an unmistakable scent, such as he had already often noticed in the shallows and amid the grass.

On several occasions, at dusk, Brighteye, like Lutra the otter, had seen a trout splashing and twisting convulsively in terror and pain. Each time the trout had been irresistibly drawn through the shallows towards a peculiar, upright object on the opposite bank, and after this object had passed into the distance the vole had found that the familiar scent of which he was now conscious was mingled, at the edge of the river-bank, with fresh blood-stains and with the strong smell of fish.

To all animals, whether wild or domesticated, fresh-spilt blood has a significance that can never be disregarded. It indicates suffering and death. Ever since, in far distant years, blood first welled from a stricken creature's wounds, Nature has been haunted by the grim presence of Fear. The hunting weasel, coming unexpectedly to a pool of blood, whence a wounded rabbit has crawled away to die in the nearest burrow, opens mouth and nostrils wide to inhale [Pg 76] with fierce delight the pungent odour. Once I caught sight of a weasel under such circumstances, and was startled by the almost demon-like look of ferocity on the creature's face.

But the hunted weaklings of the fields and woods read the signs of death with consternation. When the scent of the slayer is mingled with that of the victim it is noted with care, and, if often detected in similar conditions, is committed to memory as inseparable from danger.

Brighteye had been repeatedly warned by his mother to avoid the presence of man, and had also learned to fear it because of his experiences with the angler and the trout. Alarmed at the approach of men and hounds, he waded out, swam straight up-stream to a tiny bay, and hid beneath a willow-root to wait till the danger had passed. He strained his ears to catch each different sound as the "thud, thud" and the patter of feet came nearer. Then the gravel rattled, a stone fell into the stream, and a shaggy spaniel poked his nose into a hole between the willow-roots. The dog drew a long, noisy breath, and barked so suddenly and loudly, and so close [Pg 77] to Brighteye's ear, that the vole involuntarily leaped from his resting place.

In full view of the spaniel, Brighteye passed deep down into the clear, unruffled pool, hurriedly using every limb, instead of only his hind-legs, and with quick strokes gained the edge of the current, where for an instant he rose to breathe before plunging deep once more and continuing his journey towards the willows on the opposite bank. As he dived for the second time, Bob saw him among the ripples, and with shrill voice headed the clamouring hounds, that, "harking forward" to his cry, rushed headlong in pursuit through shallow and pool. A stout, lichen-covered branch, weighed down at the river's edge by a mass of herbage borne thither by a recent heavy flood, occupied a corner in the dense shadow of an alder; and the vole, climbing out of the water, sat on it, and was hidden completely by the darkness from the eager hounds. But his sanctuary was soon invaded; the indefatigable terrier, guided by the tiny bubbles of scent borne down by the stream, left the river, and ran, whimpering with excitement, straight to the alder. Brighteye [Pg 78] saw him approach, dived silently, and, with a wisdom he had never gained from experience, turned in a direction quite contrary to that in which the terrier expected him to flee. The vole moved slowly, right beneath the dark form of the terrier now swimming in the backwater. On, on, he went, past the stakes at the outlet of the pool into the trout-reach, and still on, by a series of dives, each following a brief interval for breath and observation among the sheltering weeds, till he arrived at the pool above the cottage gardens, where a wide fringe of brushwood formed an impenetrable thicket and he was safe from his pursuers.

Hardly, however, was this long journey needed. The dog was baffled at the outset; and, casting about for the lost scent, he discovered, on the pebbles, the strong smell of the weasel that had wandered thither to quench his thirst while Brighteye was feeding in the reed-bed opposite. Bob never by any chance neglected the opportunity of killing a stoat or a weasel; so, abandoning all thoughts of rats and voles, he dashed upward through the wood, and, almost immediately closing on his [Pg 79] prey, destroyed a bloodthirsty little tyrant that, unknown to Brighteye, had just been planning a raid on the burrow by the willow-stoles.

Water-voles, as a rule, are silent little creatures; unless attacked or frightened they seldom squeak as they move in and out of the lush herbage by the riverside. But Brighteye was undoubtedly different from his fellows: he was almost as noisy as a shrew in the dead leaves of a tangled hedgerow, and his voice was like a shrew's, high-pitched and continuous, but louder, so that I could hear him at some distance from his favourite resort in the reeds and the rushes by the willows. He seemed to be always talking to himself or to the flowers and the river as he wandered to and fro in search of tit-bits; always debating with himself as to the chances of finding a tempting delicacy; always querulous of danger from some ravenous tyrant that might surprise him in his burrow, or pounce on him unawares from the evening sky, or rise, swift, relentless, eager, from the depths beneath him as he swam across the pool.

When I got to know him well, my [Pg 80] favourite method, in learning of his ways, was to lie in wait at a spot commanding a view of one or other of the narrow lanes joining the main road of the riverside folk, and there, my face hidden by a convenient screen of interlacing grass-stems, to listen intently for his approach. Generally, for five minutes or so before he chose to reach my hiding place, I could hear his shrill piping, now faint and intercepted by a mound, or indistinct and mingled with the swirl of the water around the stakes, then full and clear as he gained the summit of a stone or ridge and came down the winding path towards me. Though in his talkative moments Brighteye usually reminded me of the tiny shrew, there were times when he reminded me more forcibly of an eccentric mouse that, a few years before, had taken up her quarters in the wall of my study, and each night, for more than a week, when the children's hour was over and I sat in silence by my shaded lamp, had made her presence known by a bird-like solo interrupted only when the singer stayed to pick up a crumb on her way across the room. [Pg 81]

The times when Brighteye wandered, singing, singing, down the lanes and main road of the river-bank, were, however, infrequent; and the surest sign of his approach, before he came in sight, was the continuous, gossiping twitter I have already described. This habit of singing and twittering was not connected with amorous sentiments towards any sleek young female; Brighteye adopted it long before he was of an age to seek a mate, and he ceased practising his solos before the first winter set in and the morning sun glanced between leafless trees on a dark flood swirling over the reed-bed where in summer was his favourite feeding place.

Whether or not the other voles frequenting the burrow by the willows had shown their disapproval of such a habit I was never able to discover. One fact, however, seemed significant: Brighteye parted from his parents as soon as he was sufficiently alert and industrious to manage his own affairs, and, having hollowed out a plain, one-roomed dwelling, with an exit under the surface of the water and another near some primrose-roots above the level of flood, lived there for [Pg 82] months, timid and lonely, yet withal, if his singing might be regarded as the sign of a gladsome life, the happiest vole in the shadowed pool above the village gardens.

It has been supposed by certain naturalists that the song of the house-mouse is the result of a disease in its throat, and is therefore a precursor of death. The mouse that came to my study ceased her visits soon after the week had passed and was never seen again; and I was unable to determine how her end was hastened. Brighteye could not, at any rate, have suffered seriously, else he would have succumbed, either to some enemy ever ready to prey on the young, the aged, the sick, and the wounded of his tribe, or to starvation, the well-nigh inevitable follower of disease in animals. He always seemed to me to be full of vitality and happiness, as if the dangers besetting his life only provided him with wholesome excitement, and sharpened his intellect far more finely than that of the rest of his tribe.

III.

WILD HUNTING.

Once, during the first summer of the water-vole's life, I saw as pretty a bit of wild hunting as I have ever witnessed, and my pleasure was enhanced by the fact that the quarry escaped unharmed. Early in the afternoon, instead of during twilight, I, in company with the members of the village Hunt and their mongrel pack, had searched the stream and its banks for rats, and had enjoyed good sport. Suddenly, however, our ragamuffin hounds struck the line of nobler game: Lutra, the otter, was astir in the pool.

I was not surprised, for on the previous night, long after the moon had risen and sleep had descended on the village, I, with Ianto the fisherman, had passed the spot on returning from an angling expedition eight or ten miles up-stream, and had stayed awhile to watch the most expert of all river-fishers, as she dived and swam from bank to bank, and sometimes, turning swiftly into the backwater, landed on the shingle close by Brighteye's reed-bed, to devour at leisure a captured trout.

Lutra soon baffled our inexpert hounds, and gained refuge in a "strong place" well behind a fringe of alder-roots, whence Bob, notwithstanding his most strenuous efforts, failed to "bolt" her. I then drew off the hounds, led them towards the throat of the pool, and for a half hour assisted them to work the "stale drag," till I reached a bend of the river where Lutra's footprints were still visible on the fine, wet sand at the brink of a rapid.

Later, when the dogs were quietly resting at their homes, I returned, alone, to my hiding place not far from Lutra's "holt." As long as daylight lasted I saw nothing of vole or otter, though several brown rats, undeterred by the disturbance of the early afternoon, came from their burrows and ran boldly hither and thither through the arched pathways of the rank grass by the edge of the bank. The afterglow faded in the western sky around the old church beyond the village gardens; and the night, though one by one the stars were lighted overhead, became so dark that I could see nothing plainly except the white froth, in large round masses, floating idly down the pool. I waited impatiently for the moon to rise, for I feared lest the faint, occasional plashes in the pool indicated that the otter had left her "holt," and would probably be fishing in a distant pool when an opportunity for observation arrived.

The night was strangely impressive, as it always is to me while I roam through the woodlands or lie in hiding to watch the creatures that haunt the gloom-wrapt clearings among the oaks and the beeches. In the darkness, long intervals, during which nothing will be seen or heard, must of necessity be spent by the naturalist; and in such intervals the mind is often filled with what may, perhaps, be best described as the spiritual influence of night, when the eyes turn upward to the stars or to the lights of a lone farmstead twinkling through the trees, and imagination, wondering greatly at its [Pg 86] own daring, links time with eternity, and the destinies of this little world with the affairs of a limitless universe.

At length the rim of the full moon appeared above the crest of the hill behind the village, and gradually, as the orb ascended, the night became brighter, till the whole surface of the pool, except for a fleeting shadow, was clear and white, and a broad silver bar lay across the ripples between me and the reed-bed on the further side. For a time no sign of a living creature was visible; then a brown rat crept along the bank beneath my hiding place; a dim form, which from its size I concluded was that of Lutra, the otter, crossed a spit of sand about a dozen yards above the reed-bed, where a moonbeam glanced through the alders; and a big brown owl, silhouetted against the sky, flew silently up-stream, and perched on a low, bare branch of a Scotch fir beside the grass-grown path.

After another uneventful interval a slight movement was observable in the reeds directly opposite. Straight in the line of the silver bar a water-vole came towards [Pg 87] me, only the head of the little swimmer being visible at the apex of a V-shaped wake lengthening rapidly behind him. More than half-way across the pool a large boulder stood out of the water, but the vole was heading towards the bank above. Then, apparently without cause, he turned quickly and made straight for the stone. He had barely landed and run round to hide in a shallow depression of the stone when the water seemed to swell and heave immediately beside the boulder, and Lutra's head, with wide-open jaws, shot above the current. Disappointed, the otter vanished under the shining surface of the stream, came to sight once more in an eddy between the boulder and the bank, and once more disappeared. I was keenly interested, for every movement of the vole and the otter had been plainly discernible, so bright was the night, and so close were the creatures to my hiding place; and, raising myself slightly, I crawled a few inches nearer the edge of the overhanging bank.

"AN OPPORTUNITY CAME, WHICH, HAD SHE BEEN POISED IN THE AIR, COULD SCARCELY HAVE BEEN MISSED." To List

For a long time the vole, not daring to move, remained in the shadow. I had almost concluded that he had dived through [Pg 88] some crevice into the dark water on the other side of the boulder, when he cautiously lifted his head to the light, and crept into a grass-clump on the top of the stone. Thence, after a little hesitation, he moved to the edge, as if contemplating a second swim. Fastidious as to his toilet, even in the presence of danger, he rose on his haunches and washed his round, furry face. The action was almost fatal. The brown owl, that had doubtless seen him by the grass-clump and had therefore left her perch in the fir-tree, dropped like a bolt and hovered, with wings nearly touching the silver stream, above the spot where she had marked her prey. But she was too late—the vole had dived. Yet, even while, having alighted on the boulder, the owl stood baffled by the disappearance of the vole, an opportunity came, which, had she been poised in the air, could scarcely have been missed. Close to the near bank a wave rose above the surface of the eddy as Lutra, having seen the vole dive from the stone, again hurried in pursuit. So fast was the otter that the momentum carried her well into the shallows. But for the third time the vole [Pg 89] escaped. I indistinctly saw him scramble out, and run, with a shrill squeak, across a ridge of sand, offering a second chance to the listening owl; and, from his flight in the direction of the well known burrow, I concluded that the hunted creature was russet-coated little Brighteye. But the bird knew that she could not rise and swoop in time; so, probably disturbed by the presence of the otter, she flew away down-stream just as Lutra, since the vole was out of reach, glided from the sand and philosophically turned her attention to less evasive trout and eels.

Then all was motionless and silent, but for an occasional faint whistle as Lutra fished in the backwater at the throat of the pool, the wailing cry of the owl from the garden on the crest of the slope behind me, and the ceaseless, gentle ripple of the river. At last, when the voices of the otter and the owl were still, and when the shadows were foreshortened as the moon gazed coldly down between the branches of the fir, Brighteye, having recovered from his recent fright, left his sanctuary by the roots of the willow, and wandered, singing, singing, down the [Pg 90] white, winding run-way and out into the main road of the riverside people, till he came to a jutting branch above the river's brim, whence he dived into the placid pool, and swam away towards the reed-bed. Then the crossed shadows of the flags and hemlocks screened him from my sight.

The first autumn in the water-vole's life was a season of wonderful beauty. A few successive frosts chilled the sap in the trees and the bushes near the river, but were succeeded by a long period when the air was crisp yet balmy, and not a breath of wind was noticeable except by the birds and the squirrels high among the giant beeches around the old garden, and when the murmur of summer insects was never heard by night, and only by day if a chance drone-fly or humble-bee visited a surviving clump of yellow ragweed by the run-way close to Brighteye's burrow. The elms and the sycamores glowed with purple and bronze, the ash-trees and the willows paled to lemon yellow, the oaks arrayed themselves in rich and glossy olive green; while the beeches in the glade, and the brambles along the outskirts of the thickets, ruddy and golden [Pg 91] and glittering in the brief, delicious autumn days, seemed to filter and yet stain the mellow sunshine, and to fill each nook with liquid shadow as pure and glorious as the blue and amber lights on the undulating hills. Spread on the bosom of the brimming river, and broken, here and there, by creamy lines of passing foam, the reflections of this beauty seemed to well and bubble, from unfathomable deeps, around the "sly, fat fishes sailing, watching all."

The water became much colder than in summer; but Brighteye, protected by a warm covering of thick, soft fur through which the moisture could not penetrate, as well as by an over-garment of longer, coarser hair from which the drops were easily shaken when he left the stream, hardly noticed the change of temperature. But he well knew there were changes in the surroundings of his home. The flags in the reed-bed were not so succulent as they had been in early summer; the branches that sometimes guided him as he swam from place to place seemed strangely bare and grey; the clump of mayweed that, growing near his burrow, had served as a beacon in the gloom, was faded to a few [Pg 92] short brown tufts; and nightly in his wanderings he was startled by the withered leaves that, like fluttering birds, descended near him on the littered run-ways or on the glassy surface of the river-reach. It was long before he became accustomed to the falling of the leaves, and up to the time when every bough was bare the rustling flight of a great chestnut plume towards him never failed to rouse the fear first wakened by the owl, and to send him on a long, breathless dive to the bottom of the pool.

Brighteye was a familiar figure to all the river-folk, while he, in turn, knew most of them, and had learned to distinguish between friends and foes. But occasionally he made a slight mistake. Though shy, he was as curious as the squirrel that, one afternoon when Brighteye was early abroad, hopped down the run-way to make his acquaintance, and frightened him into a precipitate retreat, then ran out to a branch above the stream and loudly derided the creature apparently drowning in the stream.

An object of ceaseless curiosity to Brighteye was a water-shrew, not more than half [Pg 93] the size of the vole, that had come to dwell in the pool, and had tunnelled out a burrow in the bank above the reed-bed. Nightly, after supper, Brighteye made a circuit of the pool to find the shrew, and with his companion swam hither and thither, till, startled by some real or imagined danger, each of the playmates hurried to refuge, and was lost awhile to the other amid the darkness and the solitude of the silent hours.

Brighteye soon became aware of the fact that some of the habits of the shrew were entirely different from his own. While the vole was almost entirely a vegetable feeder, the shrew, diving to the bed of the river, would thrust his long snout between the stones, and pick up grubs and worms and leeches sheltering there. With Brighteye's curiosity was mingled not a little wonderment, for the shrew's furry coat presented a strange contrast of black above and white beneath, and, immediately after the shrew had dived, a hundred little bubbles, adhering to the ends of his hair, caused him to appear like a silvery grey phantom, gliding gracefully, though erratically, from stone to stone, from patch to patch of water-weed, from ripple to [Pg 94] ripple near the surface of the stream. The young brown trout, hovering harmlessly above the rocky shelves and in the sandy shallows, far from being a source of terror to Brighteye, fled at his approach, and seldom returned to their haunts till he had reached the far side of the current. Emboldened by the example of the shrew, that sometimes made a raid among the minnows, and desirous of keeping all intruders away from the lower entrance to his burrow, Brighteye habitually chased the trout if they ventured within the little bay before his home. But there was one trout, old and lean, whose haunt was behind a weed-covered stone at the throat of the pool, and of this hook-beaked, carnivorous creature, by which he had once been chased and bitten, Brighteye went in such constant fear that he avoided the rapid, and, directly he caught a glimpse of the long, dark form roving through the gloomy depths, paddled with utmost haste to his nearest landing place.

Since, under the care of his mother, he made his earliest visit to the reed-bed, Brighteye had seen hundreds of giant [Pg 95] salmon; the restless fish, however, did not stay long in the pool, but after a brief sojourn passed upward. Often at dusk the salmon would leap clear into the air just as Brighteye came to the surface after his first dive, and once so near was a sportive fish that the vole became confused for the moment by the sudden turmoil of the "rise," and rocked on the swell of the back-wash like a boat on the waves of a tossing sea. During the summer Brighteye had suffered nothing, beyond this one sudden fright, from the visits of the great silvery fish to the neighbourhood of his home; and, notwithstanding his experience, he was accustomed to dive boldly into the depths of the "hovers," and even to regard without fear the approach of an unusually inquisitive salmon. Late in the autumn, however, Brighteye noticed, with unaccountable misgiving, a distinct change in the appearance of these passing visitors. The silvery sheen had died away from their scales, and had been succeeded by a dark, dull red; and the fish were sluggish and ill-tempered. Besides, they were so numerous, especially after a heavy rainfall, that the stream seemed [Pg 96] barely able to afford them room in their favourite "hovers," and the old trout, previously an easy master of the situation, found it almost beyond his powers to keep trespassers from his particular haunt in mid-current at the throat of the pool. So occupied was he with this duty that he seldom roamed into the little bays beneath the alder-fringes; and Brighteye, so long as he avoided the rapid, was fairly safe from his attack. The reed-bed, though partly submerged, still yielded the vole sufficient food; and to reach it straight from his home he had to pass through the shallows, which extended for a considerable distance up-stream and down-stream from the gravelly stretch immediately outside the reeds.

About the beginning of winter, when the migration of the salmon had become intermittent, and the sea-trout had all passed upward beyond the pool, two of the big, ugly "red fish," late arrivals at the "hover" nearest the burrow, made a close inspection of the pool;

then, instead of following their kindred to the further reaches, they fell back toward the tail of the stream and [Pg 97] there remained. After the first week of their stay, Brighteye found them so ill-tempered that he dared not venture anywhere near the tail of the stream; and, as the big trout at the top of the pool showed irritation at the least disturbance, the vole was forced to wander down the bank, to a spot below the salmon, before crossing the river on his periodical journeys to the reed-bed. His kindred, still living in the burrow where he had been born, were similarly daunted; while the shrew became the object of such frequent attack—especially from the bigger of the two salmon, an old male with a sinister, pig-like countenance and a formidable array of teeth—that escape from disaster was little short of miraculous.

Having calculated to a nicety his chances of escape, and having decided to avoid at all times the haunts of the pugnacious fish, Brighteye was seldom inconvenienced, except that he had to pass further than hitherto along the bank before taking to the water, and thus had to risk attack from weasels and owls. But soon, to his dismay, he discovered that the salmon had shifted their quarters to the shallow close by the reeds. [Pg 98] He was swimming one night as usual into the quiet water by the reed-bed, and, indeed, had entered a narrow, lane-like opening among the stems, when he felt a quick, powerful movement in the water, and saw a mysterious form turn in pursuit of him, and glide swiftly away with a mighty effort that caused a wave to ripple through the reeds, while the outer stalks bent and recoiled as if from the force of a powerful blow. On the following night he was chased almost to the end of the opening among the reeds, and barely escaped; but this time he recognised his pursuer. Afterwards, having unexpectedly met the shrew, he crept with his companion along by the water's edge as far as the ford, and spent the dark hours in a strange place, till at dawn he crossed the rough water, and sought his home by a path the further part of which he had not previously explored.

IV.

SAVED BY AN ENEMY.

The days were dim and the nights long, and thick, drenching mists hung over the gloomy river. The salmon's family affairs had reached an important stage; and the "redd," furrowed in the gravel by the mated fish, contained thousands of newly deposited eggs. And, as many of the river-folk, from the big trout to the little water-shrew, continually threatened a raid on the spawn, the salmon guarded each approach to the shallows with unremitting vigilance.

It happened, unfortunately for Brighteye, that, while the construction of the "redd" was in progress, some of the eggs—unfertilised and therefore not heavy enough to sink to the bottom of the water—were borne slowly by the current to the ford below the pool, just as the shrew was occupied there in vain attempts to teach the vole how to hunt for insects among the pebbles.

If Brighteye had been at all inclined to vary his diet, he would at that moment have yielded to temptation. Everywhere around him the trout were exhibiting great eagerness, snapping up the delicacies as they drew near, and then moving forward on the scent in the direction of the "redd." The shrew joined in the quest; and Brighteye, full of curiosity, swam beside his playmate in the wake of the hungry trout. The vole found quite a shoal of fish collected near the reeds; and for a few moments he frolicked about the edge of the shallow. He could see nothing of the old male salmon, though he caught a glimpse of the female busy with her maternal duties at the top of the "redd."

After diving up-stream and along by the line of the eager trout, he rose to breathe at the surface, when, suddenly, the river seemed alive with trout scattering in every direction, a great upheaval seemed to part the water, and he himself was gripped by one of his hind-feet and dragged violently down and across to the deep "hover" near his home. The salmon had at last outwitted the vole. The current was strong, and beneath its weight Brighteye's body was bent backwards till his fore-paws rested on the salmon's head. Mad with rage and fright, he clawed and bit at the neck of his captor. Gradually his strength was giving way, and for want of air he was losing consciousness, when, like a living bolt, Lutra, the otter, to save unwittingly a life that she had erstwhile threatened, shot from the darkness of the river-bed, and fixed her teeth in the neck of the salmon scarcely more than an inch from the spot to which the vole held fast in desperation. In the struggle that ensued, and ended only when Lutra had carried her prey to shore, Brighteye, half suffocated and but faintly apprehending what had taken place, was released. Like a cork he rose to the surface, where he lay outstretched and gasping, while the current carried him swiftly to the ford, and thence to the pool beneath the village gardens. Having recovered sufficiently to paddle feebly ashore, he sat for a time in the safe shelter of a rocky ledge, [Pg 102] unnoticed by the brown rats as they wandered through the tall, withered grass-clumps high above his hiding place. At last he got the better of his sickness and fright; and, notwithstanding the continued pain of his scarred limbs, he brushed his furry coat and limped homeward just as the dawn was silvering the grey, silent pool where the lonely salmon guarded the "redd" and waited in vain for the return of her absent mate.

Brighteye took to heart his own escape from death, and for several nights moped and pined, ate little, and frequented only a part of the river-bank in proximity to his burrow. As soon, however, as the tiny scars on his leg were healed, he ventured again to the river; and for a period danger seldom threatened him. While he was unceasingly vigilant, and always ready to seek with utmost haste the safety of his home, a new desire to take precautions against the probability of attack possessed him. When, at dusk, he stole out from the upper entrance of his dwelling, he crouched on the grassy ledge at the river's brim and peered into the little bay below. If nothing stirred between the salmon "hover" and the bank, he dropped [Pg 103] quietly into the pool, inhaled a long, deep breath, dived beneath the willow-roots, and watched, through the clear depths, each moving fish or swaying stem of river-weed within the range of his vision. But not till, after several visits to his water-entrance, he was perfectly convinced of the absence of danger, did he dare to brave the passage of the pool.

The water-entrance to the vole's burrow was situated about a foot below the summer level of the river, and in a kind of buttress of gravel and soil, which, at its base, sloped abruptly inwards like an arch. This buttress jutted out at the lower corner of a little horse-shoe bay; and hereabouts, during summer, a shoal of minnows had often played, following each other in and out of every nook and cranny beneath the bank, or floating up and flashing in sun-flecked ripples faintly stirred by a breeze that wandered lightly from across the stream.

Ordinarily, Brighteye found that the hole in the perpendicular bank served its purpose well; at the slightest disturbance he could escape thither, and, safe from pursuit, climb the irregular stairway to the hollow chamber [Pg 104] above high-water mark. But it was different in times of flood. If he had to flee from the big trout, or from the otter, when the stream rushed madly past his open doorway, he found that an interval, which, however brief, was sufficient to imperil his life, must necessarily elapse before he could secure a foothold in his doorway and lift himself into the dark recess beyond.

"THE BIG TROUT, IN HIS TORPEDO-LIKE RUSH TO CUT OFF BRIGHTEYE FROM SURE REFUGE." (See p. 105). To List

Lutra had almost caught him after his adventure with the owl. He had, however, eluded the otter by diving, in the nick of time, from the stone to which he clung before the entrance, and then seeking the land. If he had been an instant later, she would have picked him off, as a bat picks a moth from a lighted window-pane, and he would never have reached the down-stream shallow. At that time the water, clearing after a summer freshet, was fairly low. Brighteye's danger in some wild winter flood would, therefore, be far greater; so, timorous from his recent experiences, and sufficiently intelligent to devise and carry out plans by which he would secure greater safety, he occupied his spare time in the lengthening nights with driving a second shaft straight [Pg 105] inward from the chamber to a roomy natural hollow among the willow-roots, and thence in devious course, to avoid embedded stones, downward to a tiny haven in the angle of the buttress far inside the archway of the bank, where the space was so confined that the otter could not possibly follow him. Even the big trout, in his torpedo-like rush to cut off Brighteye from sure refuge, utterly failed to turn, and then enter the narrow archway, in time to catch the artful vole.

The task of digging out the second tunnel was exceedingly arduous; yet, on its completion, Brighteye, taught by the changes going on around him that months of scarcity were impending, set to work again about half-way between his sleeping chamber and the upper entrance of the burrow. Here he scratched out a small, semicircular "pocket," which he filled with miscellaneous supplies—seeds of many kinds, a few beech-nuts, hazel-nuts, and acorns, as well as roots of horse-tail grass and fibrous river-weed.

He was careful, like his small relative the field-vole, and like the squirrel in the woods above the river-bank, to harvest only ripe, undamaged seeds and nuts; and in [Pg 106] making his choice he was helped by his exquisite sense of smell. He found some potatoes and carrots—so small that they had been dropped as worthless by a passing labourer on the river-path—and selected the best, leaving the others to rot among the autumn leaves. As the "pocket" was inadequate to contain his various stores, the vole used the chamber also as a granary, and slept in the warm, dry hollow by the willow-roots.

In the depth of winter, when the mist-wreaths on the stream were icy cold and brought death to the sleeping birds among the branches of the leafless alders, and when Lutra, ravenous with hunger, chased the great grey trout from his "hover," but lost him in a crevice near the stakes, Brighteye, saved from privation by his hoarded provender, seldom ventured from his home. But if the night was mild and the stars were not hidden by a cloud of mist, he would steal along his run-way to the main road of the riverside people, strip the bark from the willow-stoles, and feed contentedly on the juicy pith; while his friend, the shrew, busy in the shallows [Pg 107] near the reed-bed, searched for salmon-spawn washed from the "redd" by the turbulent flood, or for newly hatched fry no longer guarded by the lonely parent fish long since departed on her way to the distant sea.

The spirit of winter brooded over the river valley. The faint summer music of the gold-crest in the fir-tops, the sweet, flute-like solo of the meditative thrush in the darkness of the hawthorn, and the weird, continuous rattle of the goatsucker perched moveless on an oak-bough near the river-bend, were no longer heard when at dusk Brighteye left his burrow and sat, watching and listening, on the little eminence above the river's brink. Even the drone of the drowsy beetle, swinging over the ripples of the shadowed stream or from tuft to tuft of grass beside the woodland path, had ceased. But at times the cheery dipper still sang from the boulder whence the vole had dived to escape the big brown owl; and, when other birds had gone to sleep, the robin on the alder-spray and the wren among the willow-stoles piped their glad vespers to assure a [Pg 108] saddened world that presently the winter's gloom would vanish before the coming of another spring.

Like a vision of glory, which, in the first hour of some poor wanderer's sleep, serves but to mock awhile his awakened mind with recollections of a happy past, so had the Indian summer shone on Nature's tired heart, and mocked, and passed away. The last red roseleaf had fluttered silently down; the last purple sloe had fallen from its sapless stem.

A sharp November frost was succeeded by a depressing month of mist and drizzling rain. Then the heavens opened, and for day after day, and night after night, their torrents poured down the stony water-courses of the hills. The river rose beyond the highest mark of summer freshets, till the low-lying meadow above the village was converted into a lake, and Brighteye's burrow disappeared beneath the surface of a raging flood.

Gifted with a mysterious knowledge of Nature's moods—which all wild animals in some degree possess—the vole had made ready for the sudden change. On the night preceding the storm, when in the mist even [Pg 109] the faintest sounds seemed to gain in clearness and intensity, he had hollowed out for himself a temporary dwelling among the roots of a moss-grown tree on the steep slope of the wood behind the river-path, and had carried thither all his winter supplies from the granary where first they had been stored.

Brighteye was exposed to exceptional danger by his compulsory retirement from the old burrow in the river-bank. Stoats and weasels were ever on the prowl; no water-entrance afforded him immediate escape from their relentless hostilities, and he was almost as liable to panic, if pursued for any considerable distance on land, as were the rabbits living on the fringe of the gravel-pit within the heart of the silent wood. If a weasel or a stoat had entered the vole's new burrow during the period when the flood was at its highest, only the most fortunate circumstances could have saved its occupant. Even had he managed to flee to the river, his plight would still have been pitiful. Unable to find security in his former retreat, and effectually deterred by the lingering scent of his [Pg 110] pursuers from returning to his woodland haunts, Brighteye, a homeless, hungry little vagabond, at first perplexed, then risking all in search of food and rest, would inevitably have met his fate.

But neither stoat nor weasel learned of his new abode. His burrow was high and dry in the gravelly soil under the tree-trunk; and before his doorway, as far as a hollow at the river's verge, stretched a natural path of rain-washed stones on which the line of his scent could never with certainty be followed. While many of his kindred perished, Brighteye survived this period of flood; and when the waters, having cleansed each riverside dwelling, abated to their ordinary winter level, he returned to his burrow in the buttress by the stakes, and once more felt the joy of living in safety among familiar scenes.

Since the leaves had fallen, the brown rats had become fewer and still fewer along the river, and, when the flood subsided, it might have been found that none of these creatures remained in their summer haunts. They had emigrated to the rick-yard near the village inn; many of the stoats and weasels, finding [Pg 111] provender scarce, had followed in their footsteps; and Brighteye and his kindred, with the water-shrews, the moorhens, and the coots, were unmolested in their wanderings both by night and day.

The vole's favourite reed-bed was now seldom visited. Besides being inundated, it was silted so completely with gravel that to cut through the submerged stems would have been an arduous and almost impossible task. Luckily, in his journeys along the edge of the shallows during the flood, Brighteye had found a sequestered pond, near an old hedgerow dividing the wood, where tender duckweed was plentiful, and, with delicious roots of watercress, promised him abundant food. Every evening he stole through the shadows, climbed the leaf-strewn rabbit-track by the hedge, and swam across the pond from a dark spot beneath some brambles to the shelter of a gorse-bush overhanging the weeds. There he was well protected from the owl by an impenetrable prickly roof, while he could readily elude, by diving, any stray creature attacking him from land.

Winter dragged slowly on its course, and, [Pg 112] just as the first prophecy of spring was breathed by the awakening woodlands, the warm west breezes ceased to blow, and the bleak north wind moaned drearily among the trees. Night after night a sheet of ice spread and thickened from the shallows to the edge of the current, the wild ducks came down to the river from the frost-bound moors, and great flocks of geese, whistling loudly in the starlit sky, passed on their southward journey to the coast.

For the first few nights Brighteye left his chamber only when acute hunger drove him to his storehouse in the wood. Directly he had fed, he returned home, and settled once more to sleep. At last his supplies were exhausted, and he was forced to subsist almost entirely on the pith beneath the bark of the willows. The pond by the hedgerow was sealed with ice, and he suffered much from the lack of his customary food. Half-way between his sleeping chamber and its water-entrance, a floor of ice prevented ready access to the river; and, under this floor, a hollow, filled with air, was gradually formed as the river receded from the level it had reached on the first night of frost. Brighteye's [Pg 113] only approach to the outer world was, therefore, through the upper doorway. All along the margin of the pool, as far as the swift water beyond the stakes, the ice-shelf was now so high above the river that even to a large animal like the otter it offered no landing place. Only at the stakes, where the dark, cold stream flowed rapidly between two blocks of ice, could Brighteye enter or leave the river. Partly because, if he should be pursued, the swiftness of the stream was likely to lessen his chances of escape, and partly because of a vague but ever-present apprehension of danger, he avoided this spot. It was fortunate that he did so; Lutra, knowing well the ways of the riverside people, often lurked in hiding under the shelf of ice beyond the stakes, and, when she had gone from sight, the big, gaunt trout came slyly from his refuge by the boulder and resumed his tireless scrutiny of everything that passed his "hover." At last a thaw set in, and Brighteye, awakening on the second day from his noontide sleep, heard the great ice-sheet crack, and groan, and fall into the river.

When darkness came he hurried to the water's brink, and, almost reckless with [Pg 114] delight, plunged headlong into the pool. He tucked his fore-paws beneath his chin, and, with quick, free strokes of his hind-legs, dived deep to the very bottom of the backwater. Thence he made a circle of the little bay, and, floating up to the arch before his dwelling, sought the inner entrance, where, however, the ice had not yet melted. He dived once more, and gained the outer entrance in the front of the buttress, but there, also, the ice was thick and firm. He breathed the cold, damp air in the hollow beneath the ice, then glided out and swam to land. The tiny specks of dirt, which, since the frost kept him from the river, had matted his glossy fur, seemed now completely washed away, and he felt delightfully fresh and vigorous as he sat on the grass, and licked and brushed each hair into place. His toilet completed, he ran gaily up the bank to his storehouse under the tree, but only to find it empty. Not in the least disheartened, he climbed the rabbit-track, rustled over the hedge-bank to the margin of the pond, and there, as in the nights before the frost, feasted eagerly on duckweed and watercress. On the [Pg 115] following day the ice melted in the shaft below his chamber, and he was thus saved the trouble of tunnelling a third water-passage—as a ready means of escape from the otter and the big trout, as well as from a chance weasel or stoat—which, if the ice had not disappeared, he surely would have made as soon as his vigour was fully restored.

V.

Top
  [Pg 116]

THE COURAGE OF FEAR.

The dawn, with easy movement, comes across the eastern hills; the mists roll up from steaming hollows to a cloudless sky; the windows of a farm-house in the dingle gleam and sparkle with the light. So came the fair, unhesitating spring; so rolled the veil of winter's gloom away; so gleamed and sparkled with responsive greeting every tree and bush and flower in the awakened river valley. The springs and summers of our life are few, yet in each radiant dawn and sunrise they may, in brief, be found.

Filled with the restlessness of springtide life—a restlessness felt by all wild creatures, and inherited by man from far distant ages when, depending on the hunt for his sustenance, he followed the migrations of the beasts—Brighteye [Pg 117] often left his retreat much earlier in the afternoon than had been his wont, and stole along the river-paths even while the sunshine lingered on the crest of the hill and on the ripples by the stakes below the pool.

Prompted by an increasing feeling of loneliness and a strong desire that one of his kindred should share with him his comfortable home, he occupied much of his time in enlarging the upper chamber of the burrow till it formed a snug, commodious sleeping place ceiled by the twisted willow-roots; and, throwing the soil behind him down the shaft, he cleared the floor till it was smooth and level. Then he boldly sallied forth, determined to wander far in search of a mate rather than remain a bachelor. He proceeded down-stream beside the trout-reach, and for a long time his journey was in vain. He heard a faint plash on the surface of the water, and at once his little heart beat fast with mingled hope and fear; but the sound merely indicated that the last of winter's withered oak-leaves, pushed gently aside by a swelling bud, had fallen from the bough. Suddenly, from the ruined garden above him on the brow of the slope, [Pg 118] came the dread hunting cry of his old enemy, the tawny owl. Even as the first weird note struck with far-spreading resonance on the silence of the night, all longing and hope forsook the vole. Realising only that he was in a strange place far from home, and exposed to many unknown dangers, he sat as moveless as the pebbles around him, till, from a repetition of the cry, he learned that the owl was departing into the heart of the wood. Then, silently, he journeyed onward. Further and still further—past the rocky shelf where he had landed after his escape from the salmon, and into a region honeycombed with old, deserted rat-burrows, and arched with prostrate trees and refuse borne by flood—he ventured, his fear forgotten in the strength of his desire.

Close beside the river's brink, as the shadows darkened, he found the fresh scent of a female vole. He followed it eagerly, through shallow and whirlpool and stream, to a spit of sand among some boulders, where he met, not the reward of his labour and longing, but a jealous admirer of the dainty lady he had sought to woo. After the manner of their kind in such affairs, the [Pg 119] rivals ruffled with rage, kicked and squealed as if to declare their reckless bravery, and closed in desperate battle. Their polished teeth cut deeply, and the sand was furrowed and pitted by their straining feet. Several times they paused for breath, but only to resume the fight with renewed energy. The issue was, however, at last decided. Brighteye, lying on his back, used his powerful hind-claws with such effect that, when he regained his footing, he was able, almost unresisted, to get firm hold of his tired opponent, and to thrust him, screaming with pain and baffled rage, into the pool.

The female vole had watched the combat from a recess in the bank; and, when the victor returned from the river, she crept out trustfully to meet him, and licked his soiled and ruffled fur. But for the moment Brighteye was not in a responsive mood. Though his body thrilled at the touch of her warm, soft tongue, he recognised that his first duty was to make his conquest sure. His strength had been taxed to the utmost, and, since his rage was expended and his tiny wounds were beginning to smart, he feared a second encounter and the possible loss of his lady-love. [Pg 120] So, with simulated anger, he drove her before him along the up-stream path and into the network of deserted run-ways by the trout-reach. There his mood entirely changed; and soon, in simple, happy comradeship, he led her to his home.

Brighteye was a handsome little fellow. At all times he had been careful in his toilet, but now, pardonably vain, he fastidiously occupied every moment of leisure in brushing and combing his long, fine, soft fur. Both in appearance and habits he was altogether different from the garbage-loving rat. His head was rounder and blunter than the rat's, his feet were larger and softer, and his limbs and his tail were shorter. On the under side his feet were of a pale pink colour, but on the upper side they were covered, like the field-vole's, with close, stiff hair set in regular lines from the toes to the elbows of the front limbs and to the ankles of the hind-legs, where the long, fine fur of the body took its place. A slight webbing crossed the toes of his hind-feet—so slight, indeed, that it assisted him but little in swimming—and his tiny, polished claws were plum-coloured. Except when he was listening intently for some sign of danger, [Pg 121] his small, round ears were almost concealed in his thick fur. His mate was of smaller and more delicate build—this was especially noticeable when once I saw her swim with Brighteye through the clear water beneath the bank—and she was clad in sombre brown and grey.

Household and similar duties soon began to claim attention in and around the riverside dwelling. The green grass was growing rapidly under the withered blades that arched the run-ways between the river's brink and the woodland path; and, as the voles desired to keep these run-ways clear, they assiduously cut off all encroaching stems and brushed them aside. The stems dried, and in several places formed a screen beneath which the movements of the voles were not easily discernible. Selecting the best of the dry grass-stalks, the voles carried them home, and, after much labour, varied with much consultation in which small differences of opinion evidently occurred, completed, in the sleeping chamber beneath the willow-roots, a large, round nest. The magnitude of their labour could be easily inferred from the appearance of the nest: each grass-blade [Pg 122] carried thither had been bitten into dozens of fragments, and the structure filled the entire space beyond the first of the exposed roots, though its interior, till from frequent use it changed its form, seemed hardly able to accommodate the female vole.

In this tight snuggery, at a time when the corncrake's nocturnal music was first heard in the meadow by the pool, five midget water-voles, naked and blind, were born. Brighteye listened intently to the faint, unmistakable family noises issuing therefrom, and then, like a thoughtful dry-nurse, went off to find for his mate a tender white root of horse-tail grass. For several nights he was assiduous in his attentions to the mother vole; and afterwards, his house-keeping duties being suspended, he became a vigilant sentinel, maintaining constant watch over the precious family within his home.

When the baby voles were about a week old, a large brown rat, that on several occasions in the previous year had annoyed the youthful Brighteye, returned to the pool. Wandering through the run-ways, the monster chanced to discover the opening [Pg 123] from the bank to Brighteye's chamber, and, thinking that here was a place admirably suited for a summer resort, proceeded to investigate. The vole scented him immediately, and, though the weaker animal, climbed quickly out and with tooth and nail fell upon the intruder. An instant later, the mother vole appeared, and with even greater ferocity than that of her mate joined in the keen affray in order to defend her home and family to the utmost of her powers. But the rat possessed great strength and cruel teeth, and his size and weight were such that for several minutes he successfully maintained his position. With desperate efforts, the voles endeavoured to pull the rat into the water, where, as they knew, their advantage would be greater than on land. They succeeded at last in forcing him over the bank, and in the pool proceeded to punish him to such an extent—clinging to his neck by their teeth and fore-feet, while they used their hind-claws with painful effect on his body—that, dazed by their drastic methods and almost suffocated, he reluctantly gave up the struggle, and floated, gasping, down the stream. [Pg 124]

The mother vole, though she and her spouse had proved victorious, was so unsettled by the rat's incursion, that, as a cat carries her kittens, she carried each of her young in turn from their nest to a temporary refuge in a clump of brambles. Still dissatisfied, she removed them thence to a shallow depression beside one of the runways, where, throughout the night, she nursed them tenderly. At daybreak she took them back to the warmth and the comfort of the nest. Shortly afterwards, when their eyes were opened and they were following the parent voles on one of their customary night excursions, the mother found herself face to face with a far more formidable antagonist than the rat.

The baby voles, like the offspring of nearly all land animals that have gradually become aquatic in their habits, were at first strangely averse from entering the water, and had to be taken by their parents into the pool. There the anxious mother, firm yet gentle in her system of education, watched their every movement, and encouraged them to follow her about the backwaters and shallows near their home. [Pg 125] But if either of them showed the faintest sign of fatigue, the mother dived quietly and lifted the tired nursling to the surface.

Late one evening, while the parent voles were busy with their work of family training, the old cannibal trout suddenly appeared, rose quickly at one of the youngsters swimming near the edge of the current, but, through a slight miscalculation, failed to clutch his prize. The mother vole, ever on the alert, plunged down, and, heedless of danger, darted towards her enemy. For a second or two she manœuvred to obtain a grip, then, as she turned to avoid attack, the jaws of the trout opened wide, and, like a steel trap, closed firmly on her tail. Maddened with rage and pain, she raised herself quickly, clutched at the back of her assailant, and buried her sharp, adze-shaped teeth—that could strip a piece of willow-bark as neatly as could a highly tempered tool of steel—in the flesh behind his gills. So sure and speedy was her action, that she showed no sign of fatigue when she reached the surface of the water, and the trout, his spinal column severed just behind his gills, drifted lifelessly away. [Pg 126]

Though the young voles, in the tunnelled buttress of the riverbank, lived under the care of experienced parents ever ready and resolute in their defence, and became as shy and furtive as the wood-mice dwelling in the hollows of the hedge beside the pond, they were not always favoured by fortune. The weakling of the family died of disease; another of the youngsters, foraging alone in the wood, was killed by a bloodthirsty weasel; while a third, diving to pick up a root of water-weed, was caught by the neck in the fork of a submerged branch, and drowned.

During the autumn and the winter the survivors remained with their parents; the burrow was enlarged and improved by the addition of new granaries for winter supplies, new water-entrances to facilitate escape in times of panic, and a new, commodious sleeping chamber, strewn with hay and withered reeds, at the end of a long tunnel extending almost directly beneath the river-path. The supplies in the granaries were, however, hardly needed: the winter was exceptionally mild, and the voles were generally able to obtain duckweed and watercress for food. Often, on [Pg 127] my way to the ruined garden, I noticed their footprints—indistinctly outlined on the gravel, but deep and triangular where the creatures climbed through soft and yielding soil—along the path leading to the pond in the pasture near the wood.

When spring came once more, and the scented primroses gleamed faintly in the gloom beside the upper entrance to the burrow, and the corncrake, babbling loudly, wandered through the growing grass at the foot of the meadow-hedge, the household of the voles was broken up. The young ones found partners, and, in homes not far from the burrow by the willow-stoles, settled down to the usual life of the vole, a life of happiness and yet of peril.

For still another year Brighteye's presence was familiar to me. I often watched him as he sat at the water's edge above the buttress, or on the stone in mid-stream, or on the half-submerged root of a tree washed into an angle of the pool above the stakes, and as, after his usual toilet observances, he swam thence across the reed-bed opposite the "hover" where, in autumn, the breeding salmon lurked. [Pg 128]

Then, for many months, I lived far from the well loved village. But one winter evening, after a long journey, I returned. The snow, falling rapidly, blotted out the prospect of the silent hills. The village seemed asleep; the shops were closed for the weekly holiday; not a footfall could be heard, not even a dog could be seen, down the long vista of the straggling street. The white walls of the cottages, and the white snow-drifts banked beside the irregular pavements, were in complete contrast to the radiant summer scene on which my eyes had lingered when I left the village. My feeling of cheerlessness was not dispelled even by the warmth and comfort of the little inn. Oppressed by the evidences of change, which in my disappointment were, no doubt, much exaggerated, I left the inn, and, heedless of the piercing cold and the driving snow, made my way towards the river. As I approached the stakes below the pool, a golden-eye duck rose from beside the bank, and on whistling wings flew swiftly into the gloom. I crouched in the shelter of a holly tree, and waited and watched till the cold became unendurable; [Pg 129] but no other sign of life was visible; the pool was deserted.

In summer I returned home to stay, and then, as of old, I often wandered by the river. Evening after evening, till long after the last red glow had faded from the western hill-tops, I lingered by the pool. The owl sailed slowly past; the goatsucker hawked for moths about the oaks; the trout rose to the incautious flies; the corncrake babbled loudly in the long, lush meadow grass. A family of voles swam in and out of the shallows opposite my hiding place; but none of the little animals approached the buttress near the stakes. Frequently I saw their footprints on the sandy margin, but never the footprints of Brighteye. Somehow, somewhere, he had met relentless fate.

THE FIELD-VOLE.

[Pg 131]

I.

HIDDEN PATHWAYS IN THE GRASS.

The sun had set, the evening was calm, and a mist hung over the countryside when a field-vole appeared at the mouth of his burrow in a mossy pasture. The little grey creature was one of the most timorous of the feeble folk dwelling in the pleasant wilderness of the Valley of Olwen. His life, like that of Brighteye, the water-vole, was beset with enemies; but Nature had given to him, as to the water-vole, acute senses of sight, and smell, and hearing, and a great power of quick and intelligent action. He had lived four years, survived a hundred dangers, and reared twenty healthy families; and his wits were so finely sharpened that he was recognised by a flourishing colony, which had gradually increased around his moss-roofed home, as the wisest and most wide-awake field-vole that ever nibbled a turnip or harvested a seed.

For a moment the vole sat in the mouth of the burrow, with nothing of himself visible but a blunt little snout twitching as he sniffed the air, and two beady eyes moving restlessly as he peered into the sky. Suddenly he leaped out and squatted beside the nearest stone. A robin, disturbed in his roosting place by another of his kind, flew from the hedge in furious pursuit of the intruder, and passed within a few inches of the burrow. The vole, alarmed by the rush of wings, instantly vanished; but soon, convinced that no cause for fear existed, he again left his burrow and for several minutes sat motionless by the stone.

He was not, however, idle—a field-vole is never idle save when he sleeps—but he was puzzled by the different sounds and scents and sights around him; they had become entangled, and while he watched and listened his mind was trying to pick out a thread of meaning here and there. What was the cause of that angry chatter, loud, prolonged, insistent, in the fir plantation at the bottom of the field? Some unwelcome creature, bent on mischief—

perhaps a weasel or a cat—was wandering through the undergrowth, and the blackbirds, joined by the finches, the wrens, and the tits, were endeavouring to drive it from the neighbourhood. Gradually the noisy birds followed the intruder to the far end of the slope; then, returning to their roosting places, they squabbled for the choice of sheltered perches among the ivied boughs. Silence fell on upland and valley; and the creatures of the night crept forth from bank and hedgerow, and the thickets of the wood, to play and feed under the friendly protection of the fast-gathering gloom. But the field-vole would not venture from his lair beside the stone.

A convenient tunnel, arched with grass-bents, led thither from the burrow, the post of observation being shaped through frequent use into an oval "form." The vole, though anxious to begin his search for food, was not satisfied that the way was clear to the margin of the fir plantation, for the air was infused with many odours, some so strong and new that he could easily have [Pg 136] followed their lines, but others so faint and old that their direction and identity were alike uncertain. From the signs that were fresh the vole learned the story of field-life for the day. Horses, men, and hounds had hurried by in the early morning, and with their scent was mingled that of a fleeing fox. Later, the farmer and his dog had passed along the hedge, a carrion crow had fed on a scrap of refuse not a yard from the stone, and a covey of partridges had "dusted" in the soft soil before leaving the pasture by a gap beside a clump of furze. Blackbirds, thrushes, yellow-hammers, and larks had wandered by in the grass, a wood-pigeon and a squirrel had loitered among the acorns under the oak, and a hedgehog had led her young through the briars. Rabbits, too, had left their trails in the clover, and a red bank-vole had strayed near the boundaries of the field-vole's colony. Their signs were familiar to the vole from experience; he detected them and singled them out from the old trails with a sense even truer than that of the hounds as they galloped past in the morning's chase.

There was one distinct scent, however, that [Pg 137] baffled him. At first he believed it to be that of a weasel, but it lacked the pungent strength inseparable from the scent of a full-grown "vear."

Gathering courage as the darkness deepened, the field-vole rustled from his lair, ran quickly down the slope, and crept through a wattled opening into the wood. He found some fallen hawthorn berries among the hyacinth leaves that carpeted the ground, and of these he made a hasty meal, sitting on his haunches, and holding his food in his fore-paws as he gnawed the firm, succulent flesh about the kernel of the seed. Then, with a swift patter of tiny feet on the leaf-mould, he ran down to a rill trickling over a gravelly bed towards the brook, stooped at the edge of a dark pool in the shadow of a stone, and lapped the cool, clear water. Thence he made for the edge of the wood, to visit a colony of his tribe which in spring had migrated from the burrows in the uplands. Half-way on his journey, he again suddenly crossed the line of the unknown scent, now mingled with the almost overpowering smell of a full-grown weasel. The mystery was explained: the strange trail in [Pg 138] the upland meadow had evidently been that of a young "vear" passing by the hedge to join its parent in the wood.

For a moment the vole stood petrified with terror; then he sank to the earth, and lay as still as the dead leaves beneath him. But there was no time to be lost; the "vears" were returning on their trail. In an agony of fear the mouse turned back towards his home. He ran slowly, for his limbs almost refused their office of bearing him from danger. Reaching the mouth of his burrow with great difficulty, he dropped headlong down a shallow shaft leading to one of the numerous galleries. Then, lo! his mood immediately changed; his reasoning powers became strong and clear; his parental instincts whispered that his family, like himself, was in peril. Squeaking all the while, he raced down one tunnel, then down another, turned a sharp corner beneath an archway formed by the roots of a tree that had long ago been felled; and there, in a dry nest of hay and straw, he found his mate with her helpless little family of six blind, semi-transparent sucklings only three days old. He heard on every [Pg 139] side the quick scamper of feet as, alarmed by his cries, the voles inhabiting the side passages of the burrow scurried hither and thither in wild efforts to remove their young to some imagined place of safety.

"SHE WAS HOLDING ONE OF HER OFFSPRING BY THE NECK, IN PREPARATION FOR FLIGHT." (See p. 139). To List

His mate, like her neighbours, had already taken alarm. At the moment of his arrival she was holding one of her offspring by the neck, in preparation for flight. The next instant an ominous hiss reverberated along the hollow passages; the mother vole, with her suckling, vanished in the darkness of the winding gallery; and the weasels descended into the labyrinth of tunnels hollowed out beneath the moss.

Again an almost overwhelming fear possessed the hunted vole, his limbs stiffened, his condition seemed helpless. He crawled slowly hither and thither, now passing some fellow-creature huddled in the corner of a blind alley; now lifting himself above ground to seek refuge in another part of the burrow; now pausing to listen to cries of pain which indicated how thoroughly the "vears" were fulfilling their gruesome work. It seemed that the whole colony of voles was being exterminated. [Pg 140]

Bewildered, after an hour of unmitigated dread, he quitted the place of slaughter, where every nook and corner reeked of blood or of the weasels' scent, and limped through the grass towards the hedge. In a hollow among the scattered stones he stayed till terror no longer benumbed him, and he could summon courage to seek an early meal in the root-field beyond the pasture. Directly the day began to dawn, he cautiously returned to his burrow. Though numerous traces of the havoc of the night remained, he knew, from the staleness of the weasels' scent, that his foes had departed.

At noon his mate came again to her nest, and searched for her missing offspring. But the taint of blood on the floor of the chamber told her only too well that henceforth her mothering care would be needed solely by the young mouse that she had rescued in her flight. The day passed uneventfully; the weasels did not repeat their visit. At nightfall the mother mouse, stealing into the wood, found both her enemies caught in rabbit-traps set beside the "runs" among the hawthorns. [Pg 141]

For a while peace reigned in the underground dwellings of the mossy pasture, and the young field-vole thrived amazingly; from the very outset fortune favoured him above the rest of his species. After the wholesale destruction that had taken place, little risk of overcrowding and its attendant evils remained, and, for the lucky mice surviving the raid, food was plentiful, even when later, in winter, they were awakened by some warm, bright day, and hunger, long sustained, had made them ravenous. Kweek, having no brother or sister to share his birthright, was fed and trained in a manner that otherwise would have been impossible, while his parents were particularly strong and healthy. These circumstances undoubtedly combined to make him what he eventually became—quick to form an opinion and to act, and able, once he was fully grown, to meet in fight all rivals for the possession of any sleek young she-vole he happened to have chosen for his mate.

Soon after his eyes were open, the adult voles of the colony began to harvest their winter supplies. Seeds of all kinds were [Pg 142] stored in shallow hiding places—under stones, or under fallen branches—or in certain chambers of the burrow set apart for that especial purpose; and as each granary was filled its entrance was securely stopped by a mound of earth thrown up by the busy harvesters.

The first solid food Kweek tasted was the black, glossy seed of a columbine, which his mother, busily collecting provender, chanced to drop near him as she hurried to her storehouse. Earlier in the night, just outside the burrow, he had watched her with great curiosity as she daintily nibbled a grain of wheat brought from a gateway where the laden waggons had passed. He had loitered near, searching among the grass-roots for some fragment he supposed his mother to have left behind, but he found only a rough, prickly husk, that stuck beneath his tongue, nearly choked him, and drove him frantic with irritation, till, after much violent shaking and twitching, and rubbing his throat and muzzle with his fore-paws, he managed to get rid of the objectionable morsel. Something, however, in the taste of the husk so aroused his [Pg 143] appetite for solid food, that when his mother dropped the columbine seed he at once picked it up in his fore-paws, and, stripping off the hard, glossy covering, devoured it with the keen relish of a new hunger that as yet he could not entirely understand. His growth, directly he learned to feed on the seeds his mother showed him, and to forage a little for himself, was more rapid than before. Nature seemed in a hurry to make him strong and fat, that he might be able to endure the cold and privation of winter.

By the end of November, when at night the first rime-frosts lay on the fallow, and the voles, disliking the chill mists, seldom left their burrow, Kweek was already bigger than his dam. He was, in fact, the equal of his sire in bone and length, but he was loose-limbed and had not filled out to those exact proportions which, among voles as among all other wildlings of the field, make for perfect symmetry, grace, and stamina, and come only with maturity and the first love season.

When about two months old, Kweek, for the first time since the weasels had visited [Pg 144] the burrow, experienced a narrow escape from death. The night was mild and bright, and the vole was busy in the littered loam of the hedgerow, where, during the afternoon, a blackbird had scratched the leaves away and left some ripe haws exposed to view. Suddenly he heard a loud, mocking call, apparently coming from the direction of the moon: "Whoo-hoo! Whoo-hoo-o-o-o!" It was a strangely bewildering sound; so the vole squatted among the leaves and listened anxiously, every sense alert to catch the meaning of the weird, foreboding voice. "Whoo-hoo! Whoo-hoo-o-o-o!"—again, from directly overhead, the cry rang out into the night. A low squeak of warning, uttered by the father vole as he dived into his burrow, caused the young mice foraging in the undergrowth to bolt helter-skelter towards home. Kweek, joining in the general panic, rushed across the field, and had almost disappeared underground when he felt the earth and the loose pebbles falling over him, and at the same time experienced a sharp thrill of pain. Fortunately, his speed saved him—but only by an inch. The claws of [Pg 145] the great brown owl, shutting like a vice as the bird "stooped" on her prey, laid hold of nothing but earth and grass, though one keen talon cut the vole's tail as with a knife, so that the little creature squealed lustily as he ran along the gallery to seek solace from his mother's companionship in the central chamber beyond. Yet even there he was not allowed to remain in peace. Maddened by the scent of a few drops of blood coming from his wound, the adult voles chased him from the burrow, and drove him out into the field. Luckily for him the brown owl had meanwhile flown away with another young vole in her claws. Kweek remained in safety under the hawthorns till the grey dawn flushed the southeast sky; then, his injured tail having ceased to bleed, he ventured without fear among his kindred as they lay huddled asleep in the recesses of their underground abode.

The year drew to its close, the weather became colder, and an irresistible desire for long-continued rest took possession of Kweek. His appetite was more easily satisfied than hitherto; hour after hour, by night as well [Pg 146] as by day, he drowsed in the snug corner where lay the remains of the nest in which he had been born. Winter, weary and monotonous to most of the wildlings of the field, passed quickly over his head. Scarce-broken sleep and forgetfulness, when skies are grey and tempests rage—such are Nature's gifts to the snake, the bee, and the flower, as well as to the squirrel in the wood and the vole in the burrow beneath the moss. Occasionally, it is true, when at noon the sun was bright and spring seemed to have come to the Valley of Olwen, the snake would stir in his retreat beneath the leaves, the bee would crawl to and fro in her hidden nest, the flower would feel the stir of rising sap, the squirrel would venture forth to stretch cramped limbs by a visit to some particular storehouse—the existence of which, as one among many filled with nuts and acorns, he happened to remember—and the vole would creep to the entrance of his burrow, and sit in the welcome warmth till the sun declined and hunger sent him to his granary for a hearty meal. These brief, spring-like hours, when the golden furze blossomed in the hedge-bank [Pg 147] near the field-vole's home, and the lark, exultant, rose from the barren stubble, were, however, full of danger to Kweek if he but dared to lift his head above the opening of his burrow.

On the outskirts of the wood, in a rough, ivy-grown ridge where, years ago, some trees had been felled, a flourishing colony of bank-voles—little creatures nearly akin, and almost similar in shape and size, to the field-voles—dwelt among the roots and the undergrowth. These bank-voles, probably because they lived in a sheltered place screened from the bitter wind by a wall of gorse and pines, moved abroad in the winter days far more frequently than did the field-voles. For several years a pair of kestrels had lived in the valley, and had reared their young in a nest built on a ledge of rock above the Cerdyn brook and safe beyond the reach of marauding schoolboys. The hen-kestrel, when provender became scarce, would regularly at noon beat her way across the hill-top to the ridge where the red voles lived, and, watching and waiting, with keen eyes and ready talons, would remain in the air above the burrow as if poised at the [Pg 148] end of an invisible thread. Chiefly she was the terror of the bank-voles; but often, impatient of failure, she would slant her fans and drift towards the burrows in the mossy pasture, hoping to find that the grey voles had awakened for an hour from their winter sleep.

Once, when the breeze blew gently from the south and the sun was bright, Kweek, sitting on a grassy mound, saw a shadow rapidly approaching, and heard a sharp swish of powerful wings. Though drowsy and stiff from his winter sleep, he was roused for the moment by the imminence of danger, and, barely in time, scurried to his hole. A fortnight afterwards, when, again tempted out of doors by the mildness of the weather, the vole was peeping through an archway of matted grass, the hawk, with even greater rapidity than before, shot down from the sky. Had it not been that the long grass screened an entrance on the outskirts of the burrow, Kweek would then have met his fate. He fell, almost without knowing what was happening, straight down the shaft; and the sharp talons of the hawk touched [Pg 149] nothing but grass and earth, and the end of a tail already scarred by the claws of the owl. Next day, as, moving along the galleries to his favourite exit, the vole passed beneath the shaft, he saw, straight overhead, the shadowy wings outstretched, quivering, lifting, gliding, pausing, while beneath those spreading fans the baleful eyes gleamed yellow in the slant of the south-west sun, and the cruel claws, indrawn against the keel-shaped breast, were clenched in readiness for the deadly "stoop." Fascinated, the vole stayed awhile to look at the hovering hawk. Then, as the bird passed from the line of sight, he continued his way along the underground passage to the spot where he usually left his home by one of the narrow, clean-cut holes which, in a field-vole's burrow, seem to serve a somewhat similar purpose to that of the "bolts" in a rabbit's warren; and there he again looked out. The hawk still hovered in the calm winter air, so Kweek did not venture that day to bask in the sun outside his door. As soon as he had fed, and shaken every speck of loose loam from his fur, and washed himself clean with his tiny [Pg 150] red tongue, he once more sought his cosy corner and fell asleep.

Presently a pink and purple sunset faded in the gloom of night, and a heavy frost, beginning a month of bitter cold, lay over the fields. In continuous slumber Kweek passed that dreary month, till the daisies peeped in the grass, the snowdrops and the daffodils thrust forth their sword-shaped leaves above the water-meadows, and the earliest violet unfolded its petals by the pathway in the woods.

II.

[Pg 151]
Top

THE VALLEY OF OLWEN.

Eastward, the sky was covered with pale cobalt; and in the midst of the far-spreading blue hung a white and crimson cloud, like a puff of bright-stained vapour blown up above the rim of the world. Westward, the sky was coloured with brilliant primrose; and on the edge of the distant moorlands lay a great bank of mist, rainbow-tinted with deep violet, and rose, and orange. For a space immediately on each side of the mist the primrose deepened into daffodil—a chaste yet intense splendour that seemed to stretch into infinite distances and overlap the sharply defined ridges of the dark horizon. The green of the upland pasture and the brown of the plough-land beyond were veiled by a shimmering twilight haze, in which the [Pg 152] varied tints of the sky harmoniously blended, till the umber and indigo shadows of night loomed over the hills, and the daffodil flame flickered and vanished over the last red ember of the afterglow. Thus the first calm day of early spring drew to its close.

Kweek, the little field-vole, asleep in his hidden nest beneath the moss, was roused by the promise that Olwen, the White-footed, who had come to her own beautiful valley among our western hills, whispered as she passed along the slope above the mill-dam in the glen. He uncurled himself on the litter of withered grass-bents that formed his winter couch, crept towards the nearest bolt-hole of his burrow, and peeped at the fleecy clouds as they wandered idly overhead. He inhaled long, deep breaths of the fresh, warm air; then, conscious of new, increasing strength, he continued his way underground to the granary in which, some months ago, his mother had stored the columbine seeds. But the earth had been scratched away from the storehouse door, and nothing remained of the winter supplies. Hungry and thirsty, yet not daring to roam abroad while the sun was high, the vole [Pg 153] moved from chamber to chamber of his burrow, washed himself thoroughly from the tip of his nose to the tip of his tail, then, feeling lonely, awakened his parents from their heavy sleep, and spent the afternoon thinking and dreaming, till the sun sank low in the glory of the aureolin sky, and the robin's vesper trilled wistfully from the hawthorns on the fringe of the shadowed wood. Becoming venturesome with the near approach of night, but still remembering the danger that had threatened him before the last period of his winter sleep, he lifted himself warily above the ground, and for a little while stayed near the mound of earth beside the door of his burrow. Cramped from long disuse, every muscle in his body seemed in need of vigorous exertion, while with each succeeding breath of the cool twilight air his hunger and thirst increased.

Determined to find food and water, Kweek started towards the copse. No beaten pathway guided his footsteps; wind and rain, frost and thaw, and the new, slow growth of the grass, had obliterated every trail. But by following the scent of the parent voles that had already stolen into [Pg 154] the wood, he reached in safety the banks of the rill. Having quenched his thirst, he scratched the soft soil from beneath a stone and satisfied his hunger with some succulent sprouts of herbage there exposed to sight. Soon, tired from his unwonted exertion, and feeling great pain through having torn the pads of his feet—which, like those of all hibernating animals, had become extremely tender from want of exercise—he crept home to his burrow, and rested till the soreness had gone from his limbs, and he felt active and hungry again.

For the vole, guided as he was by his appetite, the most wholesome vegetable food was a ripe, well-flavoured seed. It contained all that the plant could give; leaf and stalk were tasteless compared with it, and were accepted only as a change of diet, or as a medicine, or as a last resource. Next to a seed, he loved a tender root, or a stem that had not yet thrust itself through the soil, and was therefore crisp and dainty to the taste. But the vole did not subsist entirely on vegetable food. Occasionally, when the nights were warm, he surprised some little insect hiding in the [Pg 155] moss, and pounced on his prey almost as greedily as the trout in the stream below the hill rose to a passing fly. And just as the cattle in the distant farm throve on grain and oil-cake, and the pheasant in the copse near by on wood-ants[1] "eggs," and the trout in the Cerdyn brook on ephemerals hatched at the margin of the pool, so Kweek, the field-vole, abroad in the nights of summer, grew sleek and well conditioned on good supplies of seeds and grubs. But now, worn out by long privation, he was tired and weak.

Gradually, from the bed of winter death, from the rotting leaf-mould and the cold, damp earth, the fresh, bright forms of spring arose. The purple and crimson trails of the periwinkle lengthened over the stones; then the spear-shaped buds, prompted by the flow of pulsing sap, lifted themselves above the glossy leaves and burst into flowers. The dandelion and the celandine peeped from the grass; the primrose garlanded each sunny mound on the margin of the wood; and the willow catkins, clothed with silver and pearly grey, waved in the moist, warm breeze as it [Pg 156] wandered by the brook. The queen-ant, aroused by the increasing warmth, carried her offspring from the deep recess where, in her tunnelled nest, she had brooded over them while the north-east wind blew through the leafless boughs, and laid them side by side in a roomy chamber immediately beneath the stone that screened the spot to which, in the autumn dusk, the father vole resorted that he might watch and wait before the darkness deepened on the fields and woods. The bees from the hives in the farm garden, and innumerable flies from their winter retreats in the hedgerows, came eagerly to the golden blossoms of the furze near the bank-voles' colony. The bees alighted with care on the lower petals of the flowers, and thence climbed quickly to the hidden sweets; but the flies, heedless adventurers, dropped haphazard among the sprays, and were content to filch the specks of pollen dust and the tiny drops of nectar scattered by the honey-bees. A spirit of restlessness, of strife, of strange, unsatisfied desire, possessed all Nature's children; it raised the primrose [Pg 157] from amid the deep-veined leaves close-pressed on the carpet of the grass, it tuned the carols of the robin and the thrush, it caused the wild jack-hare to roam by daylight along paths which hitherto he had not followed save by night. Kweek felt the subtle influence; long before dark he would venture from his home, steal through the "creeps," which had now become evident because of frequent "traffic," and visit the distant colonies of his kindred beyond the wood.

Of the flourishing community living in the burrow before the weasels' raid none survived but Kweek and his parents. One night, however, the father vole, while foraging near the hedgerow, was snapped up and eaten by the big brown owl from the beech-wood across the valley. In the woodlands the greatest expert on the ways of voles was the brown owl. His noiseless wings never gave the slightest alarm, and never interfered with his sense of hearing—so acute that the faint rustle of a leaf or a grass-blade brought him, like a bolt, from the sky, to hover close to the earth, eager, inquisitive, merciless, till a movement [Pg 158] on the part of his quarry sealed its doom.

The mother vole, feeling lonely and more than ever afraid, wandered far away, and found another mate in a sleek, bright-eyed little creature inhabiting a roomy chamber excavated in the loose soil around a heap of stones on the crest of the hill. Kweek, nevertheless, remained faithful to the place of his birth. Though most of his time was spent near the colony beyond the wood, he invariably returned to sleep on the shapeless litter which was all that now remained of the neat, round nest in which he had been nursed.

Kweek's frequent visits to his kindred beyond the wood led to numerous adventures. Every member of the colony seemed suddenly to have turned to the consideration of household affairs, and a lively widow-vole flirted so outrageously with bachelor Kweek that, having at last fallen a victim to her persistent attentions, he was never happy save in her company. Unfortunately a big ruffian mouse also succumbed to the widow's wiles, and Kweek found himself awkwardly placed. He fought long and stubbornly against his [Pg 159] rival, but, unequally matched and sorely scratched and bitten, was at last forced to rustle away in the direction of his burrow as quickly as his little feet could carry him. He slept off the effects of his exhaustion and the loss of a little blood and fur, then returned, stealthily, to his well-known trysting place, but found, alas! that his fickle lady-love had already regarded with favour the charms of the enemy. Kweek caught a glimpse of her as she carried wisps of withered grass to a hole in the middle of the burrow, and at once recognised that his first fond passion had hopelessly ended.

Fortune continued to treat him unkindly: that night, while returning homewards, he was almost frightened out of his wits by the shrieks of some little creature captured by the cruel owl, and, immediately afterwards, a rabbit, alarmed by the same ominous sounds and bolting to her warren in the wood, knocked him topsy-turvy as he crouched in hiding among the leaves. These adventures taught him salutary lessons, and henceforth the confidence of youth gave place to extreme caution; he [Pg 160] avoided the risk of lying near a rabbit's "creep," and was quick to discern the slightest sign, such as a shadowy form above the moonlit field, which might indicate the approach of the slow-winged tyrant of the night.

Among animals living in communities it is a frequent custom for a young male, if badly beaten in his first love episode by a rival, to elope with a new spouse, and seek a home at some distance from the scene of his defeat. Kweek suffered exceedingly from his disappointment; it was a shock to him that he should be bullied and hustled at the very time when his passion was strongest and every prospect in his little life seemed fair and bright.

For a time he dared not match himself against another of the older voles. But in an unimportant squabble with a mouse of his own age, he soon proved the victor, and, finding his reward in the favour of a young she-vole that had watched the quarrel from behind a grass-tuft, ran off with her at midnight to his old, deserted burrow in the pasture. After thoroughly examining the various galleries in the underground labyrinth, [Pg 161] the fastidious little pair dug out a clean, fresh chamber at right angles to the main tunnel, and, contented, began in earnest the duties of the year.

April came; and often, as he sat by his door, Kweek watched the gentle showers sweep by in tall pillars of vapour through the moonbeams falling aslant from the illumined edges of an overhanging cloud, and through the shadows stretching in long, irregular lines between the fallow and the copse; and night after night the shadows near the copse grew deeper, and still deeper, as the hawthorn leaf-buds opened to the warmth of spring.

The grass-spears lengthened; the moss spread in new, rain-jewelled velvet-pile over the pasture floor; the woodbine and the bramble trailed their tender shoots above the hedge; a leafy screen sheltered each woodland home; and even the narrow path from the field-voles' burrow to the corner of the copse led through a perfect bower of half-transparent greenery. The birds were everywhere busy with their nests in the thickets; sometimes, in the quiet evening, long after the moon had risen [Pg 162] and Kweek had ventured forth to feed, the robin and the thrush, perched on a bare ash-tree, sang their sweet solos to the sleepy fields; and, with the earliest peep of dawn, the clear, wild notes of the missel-thrush rang out over the valley from the beech-tree near the river. The rabbits extended their galleries and dug new "breeding earths" in their warren by the wood; and often, in the deep stillness of the night, the call-note of an awakened bird echoed, murmuring, among the rocks opposite the pines far down the slope.

During the past few weeks great events had happened in the new-made chamber of the field-voles' burrow. Hundreds of dry grass-bents, bleached and seasoned by the winter frosts and rains, had been collected there, with tufts of withered moss, a stray feather or two dropped from the ruined nest of a long-tailed titmouse in the furze, and a few fine, hair-like roots of polypody fern from the neighbouring thicket. And now, their nursery complete, four tiny, hairless voles, with disproportionate heads, round black eyes beneath unopened lids, wrinkled muzzles, and abbreviated tails—helpless [Pg 163] midgets in form suggestive of diminutive bull-dog puppies—lay huddled in their tight, warm bed. It was a time of great anxiety for Kweek. While his mate with maternal pride went leisurely about her duties, doing all things in order, as if she had nursed much larger families and foes were never known, he moved fussily hither and thither, visiting his offspring at frequent intervals during the night, creeping into the wood and back along his bowered path, scampering noisily down the shaft if the brown owl but happened to hoot far up in the glen, and doing a hundred things for which there was not the slightest need, and which only served to irritate and alarm the careful mother-vole.

Kweek inherited his timorous disposition from countless generations of voles that by their ceaseless watchfulness, had survived when others had been killed by birds and beasts of prey; and though, in his zeal for the welfare of his family, he often gave a false alarm, it was far better that he should be at all times prepared for the worst than that, in some unguarded instant, death should drop swiftly from the sky or crawl stealthily into his hidden home. [Pg 164]

During spring, more frequently than at any other season, death waited for him and his kindred—in the grass, in the air, in the trees along the hedge-banks, and on the summit of the rock that towered above the glen. Vermin had become unusually numerous in the valley, partly because in the mild winter their food had been sufficient, and partly because the keeper, feeble with old age, could no longer shoot and trap them with the deadly certainty that had made him famous in his younger days. Bold in the care of their young, the vermin ravaged the countryside, preying everywhere on the weak and ailing little children of Nature. But fate was indulgent to Kweek; though his kindred in the colony beyond the wood, and the bank-voles in the sheltered hollow near the pines, suffered greatly from all kinds of enemies, he and his mate still managed to escape unhurt.

One night a fox, prowling across the pasture, caught sight of Kweek as he hurried to his lair. Suspicious and crafty, Reynard paused at one of the entrances to the burrow, thrust his sharp nose as far as possible down the shaft, drew a long, deep [Pg 165] breath, and commenced to dig away the soil from the mouth of the hole. Suddenly changing his mind—perhaps because the scent was faint and he concluded that his labour would not be sufficiently repaid—he ceased his exertions and wandered off towards the hedge. Next day a carrion crow, seeing the heap of earth that lay around the hole, and shrewdly guessing it to mean a treat in store, flew down from an oak-tree, and hopped sideways towards the spot. He peered inquisitively at the opening, waddled over to another entrance, returned, and listened eagerly. Convinced that a sound of breathing came from midway between the two holes he had examined, he moved towards the spot directly above the nest, tapped it sharply with his beak, and again returned to listen near the entrance. But all his artifice was quite in vain; the voles would not bolt; they were not even inquisitive; so presently, baffled in his hopes of plunder, he moved clumsily away, stooped for an instant, and lifted himself on slow, sable pinions into the air.

The mother vole, assisted in questionable fashion by meddlesome Kweek, spent [Pg 166] several hours of the following night in repairing the damage done by the fox. She drew most of the soil back into the shaft, and then, where it accumulated in the passage beneath, made the opening towards the inner chamber slightly narrower than before. Soon, moistened and hardened by the constant "traffic" of tiny feet nearly always damp with dew, the mound of earth formed a barrier so artfully contrived that even a weasel might find it difficult to enter the gallery from the bottom of the shaft.

III.

[Pg 167]
Top

A BARREN HILLSIDE.

Living a secluded life in the pasture with his little mate, Kweek escaped the close attention paid by the "vermin" to his kindred in the colony beyond the wood. The brown owl still remembered where he dwelt, but, loath to make a special nightly journey to the spot, seldom caused him the least anxiety. She seemed to content herself with a strict watch over the bank inhabited by the red voles, and over the fields on the far side of the copse, where the grey voles, notwithstanding that they supplied her with many a delicious supper, were becoming numerous. She awaited an almost certain increase among the "small deer" of the pasture, before commencing her raids on the grey voles there. As [Pg 168] events proved, however, her patience was unrewarded.

Kweek's first experience in rearing a family ended disastrously. Two of the nurslings died a few hours after birth; one, venturing from the nest too soon in the evening, was killed by a magpie; and two, while sitting out near the hedge, were trampled to death by a flock of sheep rushing, panic-stricken, at the sight of a wandering fox. By the middle of May, when another vole family of six had arrived, the number of vermin in the valley had perceptibly diminished. The old, asthmatic keeper in charge of the Cerdyn valley died, and a younger and more energetic man from a neighbouring estate came to take his place. Eager to gain the favour of his master by providing him good sport in the coming autumn, the new keeper ranged the woods from dawn till dusk, setting pole-traps in the trees, or baiting rabbit-traps in the "creeps" of stoat or weasel, and destroying nests, as well as shooting any furred or feathered creature of questionable character. The big brown owl from the beech-grove, the kestrel [Pg 169] from the rock on the far side of the brook, the sparrow-hawk from the spinney up-stream, together with the weasels, the stoats, the cats, the jays, and the magpies—all in turn met their doom.

A pair of barn-owls from the loft in the farm suffered next. These owls were great pets at the old homestead. For many years they had lived unmolested in their gloomy retreat under the tiles, and regularly at nightfall had flown fearlessly to and fro among the outbuildings, or perched on the ruined pigeon-cote watching for the rats to leave their holes.

The farmer, less ignorant than the keeper, recognised the owls as friends, and treated them accordingly. They were his winged cats, and assisted to check the increase of a plague. Like the brown owl, they knew well the habits of the voles; but their attention was diverted by the rats and the mice at the farm, and they seldom wandered far afield except for a change of diet or to stretch wings cramped by a long summer day's seclusion. The rats, however, were far from being exterminated; and so, when a little child who was all [Pg 170] sunshine to his parents in the lonely homestead died from typhoid fever, the village doctor, fearing an epidemic, advised that the pests should be utterly destroyed. Loath to use strychnine, since he knew that in a neighbouring valley some owls had died from eating poisoned rats, the farmer sought the aid of the village poachers, who, with their terriers and ferrets, thoroughly searched the stacks and the buildings. During the hunt it was noticed that about a score of rats took refuge in a narrow chamber under the eaves. The farmer, directing operations in another part of the yard, was unaware of what had occurred. The poachers, knowing nothing of the presence of the owls, pushed a terrier through the opening beneath the rafters of the loft, and blocked the hole with the rusty blade of a disused shovel. For a few moments the quick patter of tiny feet indicated that the terrier was busily engaged with his task; then cries of rage and terror came from the imprisoned dog, while with these cries were mingled the sounds of flapping wings. When at last the poachers unstopped the hole and dragged the terrier out, they found that every rat [Pg 171] had been killed, and that the place was thickly strewn with the feathers of two dying owls.

During the rest of the summer, Kweek led a strangely peaceful life, having little to fear beyond an occasional visit from Reynard, or from an astute old magpie that, evading with apparent ease the keeper's gun and pole-traps, lived on till the late autumn, when, before a line of beaters, he broke cover over some sportsmen waiting for their driven game. As soon as the leaves began to fall and exhausted Nature longed for winter's rest, the burrow in the pasture became the scene of feverish activity. Kweek was now the proud sire of five or six healthy families, and the grand-sire of many more. Even the youngest voles were growing fat and strong; and, when the numerous members of the colony set about harvesting their winter stores, ripe, delicious seeds were plentiful everywhere along the margin of the wood.

The winter was uniformly mild, with exception of one short period of great cold which brought a thorough, healthful [Pg 172] sleep to the voles; and in the earliest days of spring, when the love-calls of chaffinches and tits were heard from almost every tree, Kweek and his tribe resumed their work and throve amazingly. Every circumstance appeared to favour their well-being. But for the fox, that sometimes crouched beside an opening to the burrow and snapped up an incautious venturer peeping above ground, a young sheep-dog, whose greatest pleasure in life seemed to be found in digging a large round hole in the centre of the burrow, and an adder, that stung a few of the weaklings to death, but found them inconveniently big for swallowing, the voles were seldom troubled. Their numbers, and those of every similar colony in the neighbourhood, increased in such a fashion that, before the following autumn, both the pasture and the near ploughland were barren wastes completely honeycombed with their dwellings. Every grass-root in the pasture was eaten up; every stalk in the cornfield was nibbled through so that the grain might be easily reached and devoured; and the root-crops—potatoes, [Pg 173] turnips, and mangolds—on the far side of the cornfield were utterly spoiled; and in the hedgerows and the copse the leaves dropped from the lifeless trees, each of which was marked by a complete ring where the bark was gnawed away close to the ground.

But capricious Nature, as if regretting the haste with which she had brought into the world her destructive little children, and desiring, even at the cost of untold suffering and the loss of countless lives, to restore the pleasant Cerdyn valley to its beauty of green fields and leafy woods, sent her twin plagues of disease and starvation among the voles, till, like the sapless leaves, they withered and died. And from far and near the hawks and the owls, the weasels, the stoats, and the foxes hastened to the scene. The keeper, at a loss to know whence they came, and not understanding the lesson he was being taught, bewailed his misfortune, but dared not stay their advent. At almost any hour of the day, five or six kestrels might be seen quartering the fields or hovering here and there among the burrows. [Pg 174] And, long before dark, the stoats and the weasels, as if knowing that, fulfilling a special mission, they were now safe from their arch-enemy, the keeper, hunted their prey through the "trash" of the hedge-banks, or in and out of the passages underground.

The farm labourers, in desperate haste, dug numerous pitfalls, wide at the bottom but narrow at the mouth, and trapped hundreds of the voles, which, maddened by hunger but unable to climb the sloping sides, attacked one another—all at last dying a miserable death. Not only did the customary enemies of the voles arrive on the scene: Nature called to her great task a number of unexpected destroyers—sea-gulls from the distant coast, a kite from a wooded island on a desolate, far-off mere, and a buzzard from a rocky fastness, rarely visited save by keepers and shepherds, near the up-country lakes. Food had gradually become scarce even for the few hundred voles that yet remained. No longer were they to be seen at play together, in little groups, during the cool, hazy twilight, that, earlier in the year, shimmered like a wonderful afterglow on the mossy pasture-floor. [Pg 175] Now their only desire was for food and sleep.

Unnoticed by a passing owl, Kweek, worn to a skeleton by sickness and privation, crawled from his burrow into the moonlight of a calm, clear autumn night, and lay in the shadow of the stone where the old male vole had watched and listened for the cruel "vear." A big blow-fly, attracted, with countless thousands of his kind, to the place of slaughter and decay, had gone to sleep on the side of the stone, and Kweek, in a last desperate effort to obtain a little food, moved forward to secure his prize; but at that moment his strength failed him, his weary limbs relaxed, and the dull, grey film of death overspread his half-closed eyes.

The owl, hearing a faint sound like the rustle of a dry grass-bent, quickly turned in her flight; then, slanting her wings, dropped to the ground, and presently, with her defenceless quarry in her talons, flew away towards the woods.

THE FOX.

[Pg 177]

I.

[Pg 179]

Top

THE LAST HUNT.

A dark and wind-swept night had fallen over the countryside when Reynard left the steep slope above the keeper's cottage, and stole through gorse and brambles towards the outskirts of the covert, where a narrow dingle, intersected by a noisy rill and thickly matted with brown bracken, divided the furze from some neighbouring pine-woods.

For months nothing had occurred to disturb the peace of his woodland home. Once, about a year ago, he had fled for his life before the hounds; and again, during the last autumn, while lying hidden in the ditch of the root-crop field above the pines, he had been surprised by two sheep-dogs that nipped him sorely before he could make good his escape. But at no other time had [Pg 180] he been in evident peril, and so, though naturally cunning and suspicious, he had grown bolder, and better acquainted with the neighbourhood of cottage and farmstead than were certain members of his family living on the opposite side of the valley, among thickets hunted regularly, where guns and spaniels might be heard from early morning till close of day.

Here and there, as the fox crept stealthily among the blackthorns and the gorse-bushes, he stopped for a moment on the scent of a rabbit; but the night was not such as to induce Bunny to remain outside her cosy burrow in the bank. He examined each "creep" in the tangled clumps along his way, and sometimes, resting on his haunches, sniffed the air and listened intently for any sign to indicate the presence of a feeding coney; but even the strongest taint was "stale," and no sound could be detected that might betray the whereabouts of any creature feeding in the grass. Disappointed, the fox turned towards the uplands and crossed the hedgerow into the nearest stubble. Louping leisurely along, he surprised and killed a sleeping lark. Further on he crossed [Pg 181] the scent of a hare, but

Puss was doubtless some distance away, feeding in a quiet corner of the root-crop field. Reynard now instinctively made for the farmyard among the pines, trusting meanwhile that luck would befriend him. Across the gap, by the side of the hedgerow, and through an open gateway, he went, seeking spoil everywhere, but finding none. With all his senses alert, he climbed the low wall around the yard, peeped into the empty cart-house, and stealthily approached an open shed. There, unluckily, the dogs were sleeping on a load of hay in the furthest corner. Careful not to arouse his foes, the fox retreated, and, passing the pond at the bottom of the yard, moved silently towards another shed, in which, as he knew from a former visit, the poultry roosted. Though the door was shut, an opening for the use of the fowls seemed to afford the possibility of success. With difficulty Reynard managed to squeeze himself in, only, however, to no purpose. Just beyond the door lay a loose coil of wire, brought home by the labourers after fencing and thrown here out of the way. The fox, fearing a trap, reluctantly [Pg 182] abandoned his project, returned to the bank by the pond, and crept down the lane to a spot where the ducks were housed in a neat shelter built in the wall. But here he found everything securely fastened. At this moment a door of the farmstead creaked loudly, the light of a lantern flooded the yard, and the baffled marauder sprang over the wall and trotted across the field towards the wood.

His pace soon slackened when he found himself free from pursuit; and before he reached the end of the meadow he had regained all his cool audacity and was busily planning a visit to the cottage at the foot of the dingle. Hardly had his thoughts turned once more to hunting when fortune favoured him. A hen from the farmyard had laid her eggs in the hedgerow bordering the wood, and was brooding over them in proud anticipation of one day leading home a healthy family, thus causing an agreeable surprise to the farmer's wife. The fox almost brushed against her as he sprang over the hedge, and she paid to the utmost the penalty of indiscretion.

After feasting royally on the eggs, the fox [Pg 183] took up the dead bird, and moved slowly away through the trees towards his home. Re-entering the covert, he was met by a prowling vixen that, in company with her four young cubs, inhabited an "earth" not many yards away. Reckless through hunger and maddened by the scent of blood, she attacked him savagely, bullied him out of the possession of the dead fowl, and bore her prize away in triumph to her den. The fox endured his ill-treatment with the submission of a Stoic—he happened to be the pugnacious vixen's mate, and the sire of her family. Soon recovering from the chastisement, he set off, and skirted the covert as far as the cottage garden. Finding the gate of the hen-coop closed, he sprang on the water-butt, climbed to the roof of the shed, and tried to enter the coop from above; but there, as at the farm, he feared a trap, and dared not creep beneath the loose wire netting overhanging the shed. As he jumped from the coop to the wall of the stye, he caught sight of several rats scampering to their holes. Lying flat on the wall, he awaited patiently their re-appearance. At last one of them ventured out beneath the door of the [Pg 184] cot, and was instantly killed. But, much to his chagrin, Reynard found the carcass a decidedly doubtful tit-bit, and so, having conveyed it gingerly to the margin of the covert, he scratched a shallow hole among the rotting leaves, and buried his prey, that, perhaps, its flavour might improve with keeping. Afterwards, till the sky lightened almost imperceptibly, and a steel-blue bar, low down beneath the clouds, first signalled the coming of day, he lay motionless among the undergrowth near a warren in the dingle. Then an unsuspecting rabbit hopped out into the grass, and Reynard, his watch rewarded, disappeared with his spoil into the wilderness of the gorse.

Dawn was breaking over the hills. Blue smoke curled up into the sky from the lodge cottage at the foot of the tree-clad slope. The door of the cottage stood wide open, and the scent of the wood-fire hung on the chill, damp air filling the narrow lane. A blackbird flew into the apple-tree overlooking the thatch, shook the moisture from his wings, and cleaned his bright orange bill on a bough. Then his full, reed-like music floated over the fields. The skylarks soared [Pg 185] above the upland pastures, and a shower of song descended to the valley out of the pearl-blue haze just lifting in a cloud from the hill-top. Presently the blackbird flew from the apple-tree to feed beside the hedge, and the larks dropped from the mist into the grass. But for the crackle of the cottage fire as the keeper busied himself with the preparation of his morning meal, and the rustle of a withered leaf as the blackbird moved to and fro in the ditch, not a sound disturbed the silence of the dawn. Soon the haze lifted, leaving the dew thick on the grass by the ditch, and on the moss and the ivy in the hedgerow bank. The larks soared once more into the sky; a robin sang wistfully in the ash; a brown wren, with many a flick of her tiny wings and many a merry curtsy, hopped in and out among the trees, trilling loudly a gleeful carol. The tits flew hither and thither, twittering to each other as they flew. The hedge-sparrows' metallic notes sounded clear amid all the varied music, as the birds, moving among the hazels and gently flirting their wings, pursued their coy mates from bough to bough. Through the raised curtain [Pg 186] of the mist the sun—a white globe hardly too brilliant to be boldly looked at—illumined the dewy fields with its faint beams, till the cloud-streaked sky became a clear expanse, and the blue and brown countryside glowed with the splendour of a perfect morning. The wind changed and freshened, so that the call of a farm labourer to his team and the constant voice of the river were distinctly heard in the level valley below the wood.

As the morning advanced, signs of unusual stir and bustle were apparent in the neighbourhood of the lodge. Messengers came and went between the cottage and the mansion at the bend of the river, or between the mansion and the distant village. The keeper appeared at his door, and, after satisfying himself that the lane seemed clean and well-kept, walked off briskly in the direction of the "big house." Scarlet-coated horsemen, and high-born maids and matrons, with all the medley of the Hunt in their train, cantered along the winding road—a mirthful, laughter-loving company. There were the General, stout and inelegant, wont to take his fences carefully, who changed his [Pg 187] weight-carrying mount thrice during the day, and liked a gateway better than a thorny hedge, and for the last fifteen years had never been in at the death; and his wife, the leader of fashion, but not yet the leader of the Hunt; the Major, an old shekarry from India, who still could ride as straight and fast as any man in the west; and his niece, the belle of the countryside, whose mettlesome hunter occasionally showed a sudden fondness for taking the bit between his teeth, and carrying his mistress, with reckless abandon, over furrow and five-barred gate and through the thickest hedgerow—anywhere, so long as he had breath and the music of the hounds allured him onward in his impetuous career.

The sun glanced between the trees as they passed the cottage door. Then came the Magistrate's Clerk, faultlessly attired, with florid face and glittering eyeglass, who, in an ambitious youth, finding his name too suggestive of plebeian blood, changed a vowel in it, and thereby gave an aristocratic flavour to the title of his partnership, and who acquired, with this new dignity, the taste for a monocle, a horse, and a good cigar. Following were the members of the [Pg 188] medley—the big butcher on his sturdy pony, the "dealer" on his black, raw-boned half-bred, the publican on his stolid old mare, farmers, drovers, after-riders, on cropped and uncropped mounts more accustomed to the slow drudgery of labour than to the rollicking, hard-going hunt; and after them the crowd on foot—village children, farm labourers, and apprentices from forge and counter. Riding side by side, and earnestly conversing, were the "vet," whose horse at the last hunt bolted and left him clinging to a bough, and the shopkeeper, whose grave attire and sober mien seemed strangely out of keeping with the bright, hilarious throng. These were soon met by the main party from the meet, and hounds and hunters sped away in the direction of the hillside covert, while the onlookers adjourned to the uplands, whence an almost uninterrupted view of the valleys for miles around might be enjoyed, and the movements of the fox and his enemies followed more closely than from the hollows beneath the woods.

Reynard, abundantly satisfied with his supper of eggs and early breakfast of rabbit, was lying asleep in a tuft of grass at the top [Pg 189] of the thicket when the huntsman passed down the dingle after the meet. Awakened by the noise that reached him from below, he arose, stretched his limbs, and listened anxiously—the clatter of hoofs seemed to fill the valley. Suddenly, from the outskirts of the wood, came the deep, sonorous note of a hound, followed by the sharp rebuke of the whipper-in; Jollity, the keen-nosed puppy, was "rioting" on the cold scent near the stream. Peering between the bushes, the fox could as yet see nothing moving in the covert, but a few minutes afterwards his sharp eye caught a glimpse of a hound leaping over the bank above the gorse, followed by another, and another, and yet another, till the place seemed alive with his foes.

Whither should he flee? The dingle was occupied; men and horses were everywhere in the lane; and the hounds were closing in above the gorse. The far side of the covert offered the only chance of escape, and thither he must hie, else the hounds, now pouring down the slope, would cut off his retreat. Quickly he threaded his way through the gorse, by paths familiar only to himself and the rabbits, till he reached the [Pg 190] bank by the willows; but, even while he ran, the full chorus of the hounds echoed from hillside to hillside, as, having "struck the line," they tore madly in pursuit. He reached the edge of the covert at a point furthest from his foes—then, as he crossed the meadow, a single red-coated horseman, standing sentinel far up the hillside, gave the "view-halloo," and over the brow of the slope streamed the main body of the Hunt.

It was at once evident to Reynard that by skirting the margin of the covert he could not for the present escape, so he headed downwind towards the opposite hill, hoping to find refuge in a well-known "earth" amid the thickets. To his surprise he found the entrance "stopped" with clods and prickly branches of gorse, and had perforce to continue his flight. Having well out-distanced his pursuers, he stayed to rest for a while near the stream that trickled by the hedgerow; then, with the horrid music of the hounds again in his ears, he turned, by a long backward cast, in the direction of his home.

But he was wholly unable to shake off his pursuers. For four long hours he was hustled from covert to covert, and hillside [Pg 191] to hillside, finding no respite, no mercy, no sanctuary. Breathless, mud-stained, footsore, and sick with fright, his draggling "brush" and lolling tongue betraying his distress, he sought at last the place he had long avoided, and, entering the mouth of the den where the vixen and her cubs were hiding, lay there, almost utterly exhausted. Some minutes elapsed, during which no sound but that of his laboured breathing, and of the tiny sucklings busy by the side of the dam, disturbed the stillness.

Suddenly, a deep-voiced hound broke through the bushes and bayed loudly before the entrance. His fellow joined him, and their foreboding clamour reverberated in the chamber. Terrified, the fox crawled slowly into the recess of the den. Presently a shaggy terrier came down the tunnel, and bit him sorely on the flank. He scarcely had the courage to turn on the aggressor; but the enraged vixen, thrusting her mate aside, quickly routed the daring intruder, and followed his retreat to the very mouth of the "earth," where she turned back, threatened by the great hounds that stood without. But even the reckless courage [Pg 192] of maternity was unavailing. Soon the noise of blows and of falling earth was heard, as the passage was gradually opened by brawny farm labourers, working with spade and pick, and assisted in their task by the eager huntsman, who ever and anon thrust a long bramble-spray into the tunnel and thus ascertained the direction of its devious course.

At last the tip of the fox's "brush" was seen amid the soil and pebbles that had fallen into the chamber. The huntsman had cut two stout hazel rods; these he now thrust into the hollow, one along either flank of the fox; then, grasping their ends firmly about the exposed tail, he drew poor Reynard from his hiding place, and thrust him, defiant to the last, and with his teeth close-locked on one of the hazel rods, into an old sack requisitioned at the nearest farm. The vixen met a similar fate, while the sleek, furry little cubs, treated with the utmost gentleness, were wrapped together in the Master's handkerchief and given to the care of an attendant.

Reynard's life was nearing its close. In the meadow behind the keeper's cottage the hounds were summoned by the huntsman's horn, and the bag was opened. The scene [Pg 193] that followed marred, for some of us at least, the beauty of the bright March morning. The vixen and her cubs were carried away, and found a new home in an artificial "earth" prepared for their reception near a distant mansion.

II.

A NEW HOME.

When the vixen recovered from the excitement and distress consequent on her capture, she found herself in a commodious, well ventilated chamber, circular in shape and slightly above the level of two low and narrow passages leading into the covert. The sack had been opened at the entrance of one of these passages, and the vixen had crawled through the darkness till, finding further retreat impossible, she had lain down, with wildly beating heart, on the floor of her hiding place.

Her senses seemed to have forsaken her. Had she dreamed? Often, during the warm, quiet days of a bygone summer, while lying curled in a cosy litter of dry grass-bents—which she had neatly arranged by turning round and round, and with her sensitive black muzzle pressing or lifting into shape each refractory twig—she had dreamed of mouse-hunting and rabbit-catching; her body had moved, her limbs twitched, her ears pricked forward, and her nostrils quivered as the delightful incidents of past expeditions were recalled. And when, with a start, she had awakened, as some venturesome rabbit frisked by her lair, or a nervous blackbird, startled by her movements, made the woodlands ring with news of his discovery, she had retained for a moment the impressions of her vivid dreams. But never in her sleep had she been haunted by such a bewildering sense of mingled dread and anger, such an awful apprehension of the presence of men and hounds, as that which had recently possessed her. Now, however, all was mysteriously tranquil; the full-toned clamour of the hounds and the sharp, snarling bark of the terriers had ceased; no longer was she confined and jostled in the stuffy, evil-smelling sack that yielded to, and yet restrained, her every frantic effort to regain liberty. Her heart still beat violently, as though at any moment it might break; and she crept back towards the entrance, where she might breathe the free, fresh air.

Suddenly she realised, to the full, that the day's bitter experiences were not a dream—the scent of the human hand remained on her brush, her fur was damp and matted with meal-dust, and, alas! her little ones were missing from her side. She was furious now; at all risks she would venture forth on the long, straight journey back towards home; her helpless cubs might still be somewhere under the bushes—perchance in sore need of warmth and food, and whining for their dam.

With every mothering instinct quickened, the vixen crept down the slanting passages in the direction of a faint moonlight glimmer beyond. Reaching the end of the tunnel, she, in her impetuosity, thrust her muzzle into a mass of prickles—the "earth" had been stopped with a branch of gorse. Baffled for the time, she returned to the central chamber; then cautiously, for her eyes and nostrils were smarting with pain, she tried the other outlet, but here, too, a gorse-bush baulked her exit. Now, however, [Pg 197] a faint, familiar scent seemed to fill the passage, some tiny creatures moved and whimpered, and, with almost savage joy, the vixen discovered her cubs, alive and unharmed, huddled together near the furze. Quickly she carried them, one by one, into the chamber; then, lying beside the little creatures, which, though blind and helpless, eagerly recognised the presence of their mother, she gathered them between her limbs, covered them with her soft, warm brush, and, in a language used only amid the woodlands, soothed and comforted them, while they nestled once more beneath her sheltering care. When she had fed them and licked them clean from every taint of human touch, and when she had shaken herself free from dust and removed from her brush the man-scent left by the huntsman's right hand while "drawing" her, she became more collected in her mind and more contented with her strange, new situation.

Leaving her cubs asleep, she moved along the passage, determined, if possible, to explore the thickets in hope of finding a young rabbit or a few field-voles wherewith [Pg 198] to satisfy her increasing hunger. The entrance was still blocked with furze, but just in the spot where she had found her cubs a couple of dead rabbits lay, and from one of these, though after much misgiving, she made a hearty meal. She endeavoured, but vainly, to dig a shallow trench in which to hide the rest of her provisions; the floor of the artificial "earth" was tiled, and only lightly covered with soil. Her efforts to scratch out a tunnel around the furze-bush proved alike unavailing, so she returned to her cubs, lay down between them and the narrow opening from the chamber, and slept.

That night and the following day were spent in drowsy imprisonment, till, towards the afternoon, the vixen began to feel the pangs of thirst and made fresh efforts to escape. As she was endeavouring to dislodge the tile nearest the furze, she heard the tramp of heavy feet and the sound of human voices.

"They be nice cubs," said the "whip" to the huntsman; "as nice a little lot as ever I clapped eyes on. If only they can give us such a doing as the old vixen gave us twice last December, they'll pass muster. [Pg 199] Them Gwyddyl Valley foxes be always reg'lar fliers. Their meat ain't got too easy-like; that's why, maybe, they're always in working order. Any road, their flags o' distress (tongues) don't flop over their grinders without the hounds trim 'em hard on a straight, burning scent." "Well, we'll give 'em a good start, whatever happens," replied the huntsman; "here's two more bunnies for the larder. If the old girl shifts her quarters, find out her new "earth," and feed her well. I shouldn't like to be near the guv'nor if the young uns turn out mangy when we hustle 'em about a bit in the autumn."

The voices ceased, the furze-bushes were removed from the tunnel entrances, a cold, steady current of air filled the chamber and the passages, and the vixen knew that a way had been made for her escape. She was not, however, so foolhardy as to venture forth while the scent of her foes remained strong in the thicket; she lingered, in spite of extreme thirst, till the shadows of evening deepened perceptibly in her underground abode.

When the vixen stole out into the grass, [Pg 200] the pale moon was brightening in the southern sky, and a solitary star glimmered faintly above the tree-tops. A thrush sang his vesper from the bare branch of an oak near by, and a blackbird, startled by the sight of a strange form squatting beside the brambles, sounded his shrill alarm and dipped across the clearing towards a clump of blackthorn bushes. As soon as she heard the blackbird's warning, the vixen vanished; but, presently reappearing, she trotted across the open space and sat beneath the thorns. For some minutes she remained motionless in the dark patch of shadow, listening intently; then, passing slowly down a narrow path, she reached a trickling streamlet that fell with constant music from stone to stone between luxuriant masses of moss and lichen; and there, at a gravelly pool among the boulders, she cautiously stooped to drink. With exceeding care, she now proceeded to make a thorough inspection of the covert. The night was so calm and bright that the rabbits were feeding everywhere on the margin of the thickets, but the vixen passed them by with nothing but a casual glance; her mind, for the present, was not concerned with hunting. After skirting the covert, she turned homewards by a pathway through the trees.

At the end of the path she paused, with head bent low and hackles ruffled along the spine—the scent of another vixen lay fresh on the ground. The peculiar taint told her a complete story. The strange vixen was soon to become a mother, and probably, in anticipation of the event, inhabited an "earth" close by. Casting about like an experienced hound, she picked up the trail, and followed it into a great tangle of heather, brambles, and fern, where the scent led, by many a devious turn, to the spreading roots of a beech, beneath which a disused rabbit warren had been prepared for the little strangers presently to be brought into the world. The dwelling place was empty.

Retracing her steps as far as the spot where first she had struck the trail, then turning sharply towards the clearing, the crafty creature hastened back to the "earth," determined to remove her cubs without delay to the newly discovered abode. One by one she bore her offspring thither, holding them gently by the loose skin about their necks, and housed them all before the dispossessed tenant returned from a slow and wearisome night's hunting. The evicted vixen, seeking to enter her home, speedily recognised that in her distressed condition she was no match for her savage, active enemy, and so, reluctantly retiring, took up her quarters in the artificial "earth."

Henceforth, through all the careless hours of infancy, till summer ended and the nights gradually lengthened towards the time of the Hunter's Moon, the stillness of the woodlands was never broken by the ominous note of the horn, or by the dread, fascinating music of the hounds in full cry. Three of the cubs grew stout and strong, but the fourth was a weakling—whether from injury at the hands of the huntsman or from some natural ailment was not to be determined. He died, and mysteriously disappeared, on the very day when the rest of the cubs first opened their eyes in the dim chamber among the roots of the beech.

Vulp was the only male member of the happy woodland family. His indulgent sisters tolerated his bouncing, familiar manners as [Pg 203] if they were born to be his playthings—he was so serious and yet so droll, so stupidly self-assertive and yet so irresistibly affectionate! He seemed to take his pleasures sadly, wearing, if such be possible to a fox, an air of melancholy disdain; and yet his beady eyes were ever on the lookout for mischief, and for the chance of a helter-skelter romp with his sisters round and round the chamber, or to the entrance of the "earth," where the sprouts of the green grass and the flowers of the golden celandine sparkled as the sunlight of the fresh spring morning flickered between the trees.

As yet, Vulp was unacquainted with the wide, free world. It seemed very wonderful and awe-inspiring, as he sat by the mouth of the tunnel in the shadow of an arching spray of polypody and, for sheer lack of something better to do, half lifted himself on his hind-legs to rub his lips against the edge of a fern, or to peep, with a feeling that his whereabouts were a secret, between the drooping fronds. His mother restrained his rashness; once, when he actually thrust his head beyond the ferns, she with a stern admonition [Pg 204] warned him of his mistake; and he promptly withdrew to her side, frightened at his own boldness, but grunting in well assumed defiance of the imagined danger from which he had fled.

This, in fact, was the first lesson learned—that a certain sign from the vixen meant "No," and that disobedience was afterwards punishable according to the unwritten laws of woodland life. Another sign that he learned to obey meant "Come." It was a low, deep note, gentle and persuasive; and directly Vulp heard it he would hasten to his mother to be not only fed but also cleansed from every particle of dirt. Such toilet operations were not always welcome to the youngsters, and were sometimes vigorously resented. But the vixen had a convincing method of dealing with any refractory member of her family; she would hold the cub firmly between her fore-feet while she continued her treatment, or administered slight, well-judged chastisement by nipping her wayward offspring in some tender spot, where, however, little harm could be the result.

The cubs were ten days old when they opened their eyes, but more than three weeks [Pg 205] passed before they were allowed beyond the threshold of their home. Then, one starlight night, their mother, having returned from hunting, awoke them, and, withholding their usual nourishment, gave the signal "Come." The obedient little family followed her along the dark passage, and ventured, close at her heels, into the grass-patch in the middle of the briar-brake. Vulp was slightly more timid than his sisters were; even at that early age he showed signs of independence and distrust. While the other cubs played "follow-my-leader" with the dam, he hung back, hesitating and afraid. Even an unusual show of affection by his mother failed to reassure him. A rabbit dodged quickly across a path, and immediately he stood rigid with fright. Hardly had he recovered before an owl flew slowly overhead. Enough! He paused, motionless, till the awful presence had disappeared; then darted, with astonishing speed, straight towards the "earth," and vanished, with a ridiculously feeble "yap" of make-believe bravado, into the darkness of the den. Confidence, however, came and increased as the days and the nights went by, till, at the close of a week's experiences, Vulp [Pg 206] was as bold in danger as either of his playmates. He learned to trust his mother implicitly, and, in her absence, became the guardian of the family when some fancied alarm brought fear. He was always last in learning his lessons; but, as if to make amends, he always profited most by the teaching.

Happy, indeed, were those hours of innocence—filled with sleep, and love, and play. Till Vulp was six weeks old, he was wholly unconscious of that ravenous hunger for flesh which was fated to make him the scourge of the woodlands. Nevertheless, his instincts were slowly developing, and so, when on a second occasion the old buck rabbit that had frightened him in the thicket bolted before his eyes across the path, the little fox bristled with rage and, but for his mother's presence, would doubtless have tried to pursue the exasperating coney. Invariably, when the night was fine, the cubs gambolled about the vixen on the close-cropped sward beyond the den, climbing over her body, pinching her ears, growling and grunting, tugging at each other's brushes, and in general behaving just as healthy, happy fox-cubs might be [Pg 207] expected to behave; while the patient, careful mother looked on approvingly—save when, uniting in one strong effort, they endeavoured to disjoint her tail by pulling it over her back—and smiled, as only a fox can smile, with eyes asquint and a single out-turned fang showing white beside the half-closed lip.

A great event occurred when the mother first brought home her prey that she might educate her youngsters in the matter of appetite and prepare them for an independent existence. The victim was an almost full-grown rabbit. Laying it down close to the entrance of the "earth," the vixen called her cubs, and instantly they rushed from the den, tumbling over each other in their haste, till they gained the spot where she was waiting. At that moment, however, they caught sight of the strange grey object in the grass, and, leaping back, bolted round to their mother's side. Then, feeling safe under her care, they cautiously advanced in a row to sniff the rabbit, and wondered, yet instinctively guessed, at the meaning of the situation. The vixen growled, and, picking up her prey, carried it to the bramble-clump. The cubs followed, making all sorts of curious noises [Pg 208] in mimicking their dam, and evincing the utmost inquisitiveness as to the reason of her unexpected conduct. Presently, having succeeded in arousing their inborn passion for flesh, the vixen resorted to a neighbouring mound, and left her offspring in possession of the dead animal, on which they immediately pounced, tooth and nail. How terribly in earnest they became, how bold and reckless in their vain attempt to demolish the subject of their wrath! Vulp fastened his needle-like teeth in the throat, and each of his sisters gripped a leg, while together they jerked, strained, scolded, and threatened, till the mother, fearing lest the commotion would betray their whereabouts to some lurking foe, rated her noisy progeny and in anger drove them away. But as soon as she had gone back to her seat among the grass-bents, the youngsters returned to their work. Anyhow, anywhere, they hurled themselves on the dead creature, sometimes biting each other for sheer lack of knowing exactly what else they should bite, and sometimes simply for the excitement of a family squabble.

At last, their unwonted exertions began to tire them; then the careful vixen, desirous [Pg 209] of bringing the lesson to its close, "broke up" her prey and divided it among her hungry children. They fed daintily, choosing from each portion no more than a morsel, and soon afterwards, exhausted by excitement and fatigue, and forgetful of their differences, were fast asleep, huddled together as usual in the roomy recess of the den. For a while the vixen remained to satisfy her hunger; then, having buried a few tit-bits of her provender, she also retired to rest; and silence brooded over the woodlands till the break of day set every nesting bird atune.

The vixen proved to be an untiring teacher, and the education of the cubs occupied a part, at least, of every night. The young foxes were growing rapidly, and accompanied their dam in her wanderings about the thickets. She never went far afield, food being easily procured at that time of year, particularly as in a certain spot additional supplies for the larder were frequently forthcoming because of the vigilance of the huntsman, whose one desire was to fit the cubs to match his hounds in the first "runs" of the coming season.

III.

THE CUB AND THE POLECAT.

The young fox's education, varied and thorough, steadily proceeded. Though the vixen-cubs were slightly quicker to learn, they were more excitable, and consequently did not benefit fully by each lesson. Vulp soon began to hunt for his own sport and profit. In the meadow above the wood he would sit motionless, his eyes fixed on the ground, till the voles came from their burrows to play beneath the grass-bents; then, with a quick rush, he would secure a victim directly its presence was betrayed by a waving stalk. With the same patience he would watch near a rabbit warren, till one of the inhabitants, hopping out to the mound before her door, gave him the sure chance of a kill. But in the wheat-fields on the slope his methods were altogether different. To capture partridges required unusual cunning and skill, and such importance did the vixen attach to this branch of her field-craft, that, before initiating her youngsters into the sport of hunting these birds at night, she instructed them diligently in the methods of following by scent, training them how to pursue the winding trail left by the larks that fed at evening near their sleeping places, or by the corncrakes that wandered babbling through the green wheat. Vulp's first attempt to capture a partridge chick resulted in failure. The vixen-cubs "fouled" the line he had patiently picked out in the ditch around the cornfield, and, "casting" haphazard through the herbage, alarmed the sleeping birds, and sent them away to a secure hiding place in the clover. But his second attempt was crowned with success, and he proudly carried his prey into a sequestered nook amid the gorse, where he enjoyed a quiet meal.

The cub was fully six months old before he knew the precise difference between stale and fresh scent, or between the scent of one creature and that of another, and how to hunt accordingly; and several years, with many dangers and hair-breadth escapes, were destined to pass before he became expert in avoiding or baffling the numerous enemies—chiefly dogs, and men, and traps—that threatened his life. And yet, during the first few months of his existence, he gained sufficient knowledge for the needs of the moment; and when August drew on towards the close of the summer, and he was three parts grown, he had so extended his nightly rambles that the "lay of the land" was familiar for miles around the covert. His outdoor existence—for now he was wont to sleep in a lair among the gorse and the bracken, instead of in the stuffy "earth"—gave him strength in abundant measure, while his scrupulously clean habits, the care with which he removed even the slightest trace of a burr from his sleek, brown coat, and the plentiful supplies of fresh food which he was able to obtain, naturally preserved him from mange and similar ailments to which carnivorous animals are always prone. For the present, indeed, life meant nothing more to him than the sheer enjoyment of vigorous health, at home by day amid the [Pg 213] grateful shadows of the bushes and the trees, or basking in the sun, and abroad at night in the cold, clear air of the dewy uplands.

Just as sportsmen occasionally meet with a run of ill-luck, when for some apparently unaccountable reason they either fail to find game, or fail to kill it, and, to intensify the annoyance, an accident occurs that leaves a bitter memory, so Vulp, during one of his long rambles over the countryside, failed entirely to find sport, and gained a decidedly unpleasant experience. If only his mother had not taught him that in a season of scarcity a weasel might reasonably be considered an article of food! One summer night, as he started on his usual prowl, the covert seemed strangely silent. With the exception of a solitary rabbit that bolted to its burrow when the young fox crossed the clearing, and another that disappeared in similar fashion when nothing more than a slight crackle of a leaf betrayed Vulp's whereabouts near a bramble-clump, every animal had apparently deserted the thickets. So, leaving his accustomed haunts, he crossed the furze-clad dingle, and watched near a large warren in the open. But there, again, [Pg 214] not a rabbit could be seen. A field-vole rustled by over the leaves; the cub made a futile effort to capture it, stood for an instant listening to its movements, then thrust his nose into the herbage in another vigorous but vain attempt; the vole, like the rabbits, had sought refuge underground.

An owl, that had frightened the cub about five months before when first he ventured outside his home, rose from the hedge, and flew slowly down the valley with a little squealing creature in her talons; she, at any rate, had not hunted in vain.

At last Vulp struck a fresh line of scent which, though particularly strong and uninviting, he took to be that of a weasel. It was mingled with the faint odour of a field-vole that, doubtless, had been pursued and carried away by its persistent enemy. The cub followed the trail, hoping to secure both hunter and victim, but it soon led him to a hole in the hedgerow, and there abruptly ceased. He was about to turn from the spot, when the eyes of the supposed weasel suddenly gleamed at the mouth of the hole, but disappeared when the presence of the cub was recognised. The fox, retreating to [Pg 215] a convenient post of observation behind a tuft of grass, settled down to await his opportunity. A few minutes elapsed, and the pursued creature came once more in sight. It appeared like a shadow against the sky, lifted its nose inquiringly, quitted the burrow, sat bolt upright for a moment, then, reassured, proceeded towards the covert on the opposite side of the path. With a single bound, the cub cleared the grass-tuft, reached out at his prey, missed his grip, bowled the animal over, and, turning rapidly, caught it across the loins instead of by the throat. Unfortunately for himself, the fox had made a slight miscalculation. With a scream of rage and pain, the polecat—for such the creature proved to be—turned on the aggressor, and instantly fastened its formidable teeth, like a steel trap, on his muzzle. Vulp had been taught that his fangs, also, were a trap from which there should be no escape, and so he held on firmly, trying meanwhile to shake the life from his victim. He pressed the polecat to the ground, and frantically endeavoured to disengage its hold by thrusting his fore-paws beneath its muzzle; but every effort alike [Pg 216] was useless. A scalding, acrid fluid emitted by the polecat caused the lips and one of the eyes of the cub to smart unbearably, and the offensive odour of the fluid grew stronger and stronger, till it became almost suffocating. At last the polecat convulsively trembled as its ribs and spine were crushed in the fox's tightening jaws, its teeth relaxed their hold, and the fight was over.

Sickened by the pungent smell, and with muzzle, lips, and right eye burning horribly from his wounds and the irritant poison, Vulp hastily dropped his prey, and ignominiously bolted from the scene of the encounter. Soon, however, he stopped; the pain in his eye seemed beyond endurance. He tried to rub away the noxious fluid with his paws, but his frantic efforts only increased the irritation by conveying the poison to his other eye and to his wounds. He rolled and sneezed and grunted in torment; he drew his muzzle and cheeks to and fro on the ground, wrestling with the great Earth-Mother for help in direst agony. He could not open his eyes; he stumbled blindly against a tree-trunk, and at last became entangled in the prickly undergrowth. This was Nature's method [Pg 217] of succour—she forced her wildling to remain quiet, in helpless exhaustion, till the pain subsided and life could once again be endured. Panting and sick, the cub lay outstretched among the thorns, while the tears flowed from his eyes and the froth hung on his lips. Presently, however, relieved by the copious discharge, he recovered his senses, and, miserably cowed, with head and brush hanging low, returned before dawn to the covert. But the vixen in fury drove the cub away; the scent still clung to him, and rendered him obnoxious even to his mother. In shame he retired to a dense "double" hedge of hawthorn, where he hid throughout the day, till he could summon sufficient courage at dusk to hunt for some dainty morsel wherewith to tempt his sickened appetite. But before taking up his position above the entrance to a rabbit warren, he drank at the brook, dipped his tainted fore-paws in the running water, and, sitting by the margin, removed from his face, as far as possible, the traces left by the previous night's conflict. Repeatedly, at all hours of the day and the night, he licked his paws and with them washed his wounded muzzle and [Pg 218] inflamed eyes; but so obstinately did the offensive odour cling to him that a fortnight elapsed before the last vestige of the nuisance disappeared. Meanwhile, he narrowly escaped the mange; and, to add to the discomfort of his wounds, he experienced, now that his mother's aid was lacking, some difficulty in obtaining sufficient fresh food.

At length he recovered, and new, downy hair clothed the wounds and the scratches on his muzzle and throat. Sleek and strong once more, he was welcomed as a penitent prodigal by the relenting vixen, and, having in the period of his solitary wanderings learned much about the habits of the woodland folk, was doubtless able to assist his mother in the future training of the vixen-cubs.

In that luckless fortnight he had acquired a taste for young pheasants, had picked up a few fat pigeon-squabs belonging to the last broods of the year, and had sampled sundry articles of diet—frogs, slugs, snails, a young hedgehog or two, and a squirrel that, overcome with inquisitiveness, descended from the tree-tops to inspect the young fox as he dozed among the bilberries carpeting the forest floor. [Pg 219]

Another incident occurred, to which, at the time, the cub attached considerable importance. He had killed what seemed to be a large, heavy rabbit, which, though evidently possessed of a healthy appetite, was almost scentless, and differed in taste from any he had hitherto captured. He was not particularly hungry, so he buried the insipid flesh, and resolved never to destroy another rabbit that did not yield a full, strong scent. Shortly afterwards, when, under the eye of the bright August moon, Vulp and the vixen were hunting in the wheat-fields, he detected a similarly weak scent along the hedgerow, and learned from his wise mother it was that of a doe-hare about to give birth to her young, and therefore hardly worth the trouble of following. The vixen further explained that, except when other food was scarce, creatures occupied, or about to be occupied, with maternal cares—even the lark in the furrow and the willow-warbler in the hole by the brook—were far less palatable than at other times. The cub was also told how, just before he came into the world, the hounds had chased his mother from the thicket, and how old Reveller, the leader [Pg 220] of the pack, had headed the reckless puppies, and, rating them for their discourtesy, had led them away to scour another part of the covert.

With the advance of autumn, a great change passed over the countryside. The young fox now found it necessary to choose his paths with care as he wandered through the darkness, lest the rabbits should be warned of his approach by the crisp rustle of his "pads" on the leaves that had fallen in showers on the grass. Hitherto he had associated the presence of man with that of something good for food. An occasional dead rabbit was still to be found near the old "earth," and, strange to relate, the man-scent leading to the place was never fresher or staler than that of the rabbit. In another spot—a wood-clearing not far from the keeper's lodge—the strong scent of pheasants always seemed to indicate that the birds had ventured thither in numbers to feed, and there, too, the man-scent was strong on the grass. The tracks of innumerable little creatures intersected the clearing in all directions, and, if but for the sport of watching the pheasants, the pigeons, the sparrows, and the voles playing and quarrelling [Pg 221] in the undergrowth or partaking of the food provided by the keeper, the fox loved to lurk in the gorse near by. He evinced little real alarm even at the sight of man, though he felt a misgiving and instinctively knew that he must hide or keep at a distance till the curiously shaped monster had gone. The vixen warned him repeatedly; and she herself, after giving the signal "Hide!" would slink away, and wander for miles before returning to her family, if only the measured footfall of a poacher or a farm labourer sounded faintly through the covert.

But soon the young fox learned, in a way not to be misunderstood, that the presence of man meant undoubted danger. One day in October, as he was intently watching the movements of a sportsman in the copse, a big cock pheasant rose with a great clatter from the brambles, a loud report rang through the covert, and a shaggy brown and white spaniel dashed yelping into the bushes. Darting impetuously from his lair, the cub easily out-distanced the dog, and quickly found refuge in an adjoining thicket, where he remained in safety during the rest of the [Pg 222] day. Night brought him another adventure. While crossing a pasture towards a wooded belt on the hillside, he discovered, to his surprise, that a man was creeping stealthily towards him through the shadows. A moment later, a great lurcher came bounding over the field. The fox turned, made for the hedgerow, and gained the friendly shelter of the hawthorns just as the dog crashed into the ditch. The frightened creature now ran along the opposite side of the hedge in a straight line towards the wood, and for a second time narrowly escaped the lurcher's teeth; but, by keeping close to the ditch and among the prickly bushes on the top of the hedge-bank, he at last succeeded in baffling his long-legged foe and reached the wood unharmed.

Vulp had thus awakened to the dangers which, during winter and the earliest days of spring, were always to beset him. But the apprehensions caused by his little affair with the spaniel, and even by his narrow escape from the lurcher, were trifling compared with the dread and distress of being driven for hours before the hounds. And so full of perils was the first winter of his life that [Pg 223] nothing but a combination of sheer luck with great endurance could then have sufficed to save him from destruction. Quickly, one after the other, the young vixens were missing from the thickets; soon afterwards, three of the cubs belonging to the litter that had been reared in the artificial "earth" disappeared; and an old fox, the sire of that litter, was killed after a long, wearisome chase almost to the cliffs on the distant coast.

One dark and dismal night in December, Vulp, on returning to the home thickets, failed to find his dam. Her trail was fresh; she had evidently escaped the day's hunt; but all his efforts to follow her met with no sort of success. Nature had brought about a separation; in the company of an adult fox, whose scent lay also on the woodland path, the vixen had departed from her haunts. The fox-cub remained, however, among the woodlands where he had learned his earliest lessons, and, for another year, hunted and was hunted — a vagrant bachelor.

IV.

[Pg 224]
Top

A CRY OF THE NIGHT.

One starlit night, when in early winter the snow lay thick on the ground, Vulp heard the hunting call of a vixen prowling through the pines. A similar call had often reached his ears. Not long after his dam deserted him, the cry had come from a furze-brake on a neighbouring hill-top, and, hastening thither, he had wandered long and wearily, recognising, though with misgiving, his mother's voice. But the exact meaning of the call, not being a matter for his mother's teaching, was unknown to him at the time. Now, however, he was a strong, well fed, fully developed fox, able to hold his own against all rivals, and the cry possessed for him a strange, new significance: "The night is white; man [Pg 225] is asleep; I hunt alone!" Almost like a big brown leaf he seemed to drift across the moonlit snow, nearer and nearer to the pines. He paused for a moment to sniff the trail; then, with a joyous "yap" of greeting, he bounded over the hedge, reached the aisles of the wood, and gambolled—again like a big, wind-blown leaf—about the sleek, handsome creature whose call he had heard. The happy pair trotted off to hunt the thickets, till, just before dawn, Vulp, eager to show his skill and training, surprised two young rabbits sitting beneath a snow-laden tangle of briar and gorse, and gallantly shared the spoil with his woodland bride. They feasted long and heartily, afterwards journeying to the banks of a rill, that, like a black ribbon, flowed through the glen; and there, crouching together at the margin, they lapped the water with eager, thirsty tongues.

Presently, happening to glance behind along the line of the trail, Vulp caught sight of another fox, a rival for the vixen's affections, crouching in some bracken scarcely a dozen yards away. With a low grunt of rage, he dashed into the fern, but the watchful [Pg 226] stranger simply moved aside, and frisked towards the vixen as she still crouched at the edge of the stream. In response to this insulting defiance, Vulp hurled himself on the intruder, and bowled him over into the snow. The fight was fast and furious; now one gained the advantage, then the other. The grass beneath them became gradually bared of snow by their frantic struggles, and marked here and there by a bunch of fur or a spot of blood. At last the rival fox, his cheek torn badly beneath the eye, showed signs of exhaustion; his breath came in quick, loud gasps; and Vulp, pressing the attack, forced him to flee for life to a thicket on the brow of the slope. There he dwelt and nursed his wounds, till, when the snow melted, the huntsman's "In-hoick, in-hoick, loo-loo-in-hoick!" resounded in the coverts, and he was routed from his lair for a last, half-hearted chase, that ended as Melody pulled him down at a ford of the river below the woods.

During the period of their comradeship—a period of privation for most of Nature's wildlings—Vulp taught the vixen much of [Pg 227] the lore he had learned from his mother, while the vixen imparted to him the knowledge she herself had gained when a cub. He taught her how to steal away from the covert along the rough, rarely trodden paths between the farm-labourers' cottages—where the scent lay so badly that the hounds were unable to follow—directly the first faint notes of a horn, or the dull thud of galloping hoofs, or the excited whimper of a "rioting" puppy, indicated the approach of enemies. She taught him to baffle his foes by chasing sheep across the stubbles, and then passing through a line of strong scent where his own trail could not readily be distinguished; also that to cross the river by leaping from stone to stone in the ford was as sure a means of eluding pursuit as to swim the pools and the shallows. He taught her, when hard pressed, to leap suddenly aside from her path, run along the top rail of a fence, return sharply on her line of scent, and follow, with a wide cast, a loop-shaped trail, which, with a tangent through a ploughed field or dry fallow, was usually sufficient to check pursuit till the scent became faint and cold. And gradually each of these woodland [Pg 228] rovers grew acquainted with the peculiar whims and habits of the other. Vulp loved to follow stealthily the trail of the rabbit, and then to lie in wait till some imagined cause of alarm sent Bunny back through the "creep" and almost straight into her enemy's open jaws. The vixen preferred to hide in the brambles to leeward of a burrow till an unsuspecting rabbit crept out into the open. Vulp, since his adventure with the polecat, bristled with rage whenever he crossed the track of a weasel, but never dreamed of following; polecat and weasel were the same animal for aught he knew to the contrary. The vixen, however, was not daunted by the unpleasant memory of any such adventure; having chanced to see a weasel in the act of killing a vole, she had recognised a rival and acted accordingly. And so Vulp's repeated warnings to his mate on this matter produced no effect beyond making her slightly more careful than she had hitherto been to obtain a proper grip when she pounced on her savage little quarry. The vixen was exceedingly fond of snails, and would eagerly thrust a fore-paw into the crannies of any old wall or bank where they hibernated; but Vulp [Pg 229] much preferred to scratch up the moss in a deserted gravel-pit, and grub in the loosened soil for the drowsy blow-flies and beetles that had chosen the spot for their winter abode. This was the reason for such different tastes: the vixen, when a cub, had often basked in the sun near a snails' favourite resort, and had there acquired a liking for the snails; while the fox, on the other hand, had times out of number amused himself, in the first summer of his life, by leaping and snapping at the flies as they buzzed among the leaves when the mid-day sun was hot, and at the beetles as they boomed along the narrow paths in the thicket near the "earth" when the moon rolled up above the hedge, and the dark, mysterious shadows of intersecting boughs foreshortened on the grass. But Vulp knew well, from an unpleasant experience, the difference between a fly and a wasp.

One day in August, as he lay in his outdoor lair, the brightness and heat of the sunshine were such that his eyes, blinking in the drowsiness of half-awakened slumber, appeared like mere slits of black across streaked orbs of yellow, and gave no indication [Pg 230] of the fiery glow that lit the round, distended pupils when he peered at nightfall through the tangled undergrowth. His tongue lolled out, and he panted like a tired hound, but from thirst rather than weariness. The flies annoyed him greatly, now settling on his brush, till with a flick of his paw he drove them away, then, nothing daunted, alighting on his back, his ears, his haunches, till his fur wrinkled and straightened in numberless uneasy movements from the tormenting tickling of the little pests. Presently, with a shrill bizz of rapid wings, a large, yellow-striped fly passed close to his ears. He struck down the tormenting insect with a random flip of his paws, snapped at it to complete the work of destruction, and proceeded leisurely to eat his victim. To his utter surprise, he seemed to have captured a living, angry thorn, which, despite his most violent efforts to tear it away with his paws, stuck in his lip, and produced a smarting, burning sensation that was intolerable. He rolled on the ground and rubbed his muzzle in the grass, but to no purpose. No wonder, then, that subsequently his manner towards [Pg 231] an occasional hibernating wasp among the moss-roots in the gravel-pit was deferential in the extreme!

Vulp and his mate soon learned that in rabbit-hunting it was exceedingly profitable to co-operate. Thus, while the vixen "lay up" near a warren, Vulp skirted the copse and chased the conies home towards his waiting spouse. After considerable practice, the trick paid handsomely, and food was seldom lacking. The vixen possessed, perhaps, a slightly more delicate sense of smell than the fox. Frequently she scented a rabbit in a clump of fern or gorse after Vulp had passed it by; suddenly stopping, she would tell her lord of her discovery by signs he readily understood, and then, while he kept outside the tangle, would pounce on the coney in its retreat, or start it helter-skelter into his very jaws. But of all the tricks and the devices she taught him, the chief, undoubtedly, were those concerned with the capture of hens and ducks from a neighbouring farmstead. An adult fox, as a rule, does not pay frequent visits to a farmstead; but Vulp, like his sire, was passionately fond of poultry, and [Pg 232] so, in after years, the vixen's instructions caused him to become the dread of every henwife in the district. Undoubtedly he would have been shot had he not been the prize most sought for by the Master of the Hounds, who cared little for the frequent demands made on his purse by the cottagers, so long as the fox that slaughtered the poultry gave abundant sport when running fast and straight before the pack.

The months drifted by, and signs of spring became more and more abundant in the valley. About the beginning of March, Vulp deserted the "earth" prepared by himself and the vixen for their prospective family, and took up his abode among the hazels and the hawthorns in a thick-set hedge bounding the woods.

In preparing the "breeding earth," Vulp and the vixen observed the utmost care in order that its whereabouts should not be discovered. The chosen site was a shallow depression, scratched in the soil by a fickle-minded rabbit that had ultimately fixed on another spot for her abiding place. This depression was enlarged; a long tunnel [Pg 233] was excavated as far as the roots of an oak, and there broadened. Then another long tunnel was hollowed out towards the surface, where it opened in the middle of a briar-brake. The foxes worked systematically, digging away the soil with their fore-paws, loosening an occasional stubborn stone or root with their teeth, and thrusting the rubbish behind them with their powerful hind-legs. As it accumulated, they turned and pushed it towards the mouth of the den, where at last a fair-sized mound was formed. When the burrow had been opened into the thicket, the crafty creatures securely "stopped" the original entrance, so that, when the grass sprouted and the briar sprays lengthened in the woodlands, the "earth" would escape all notice, unless a prying visitor penetrated the thicket and discovered the second opening—then, of course, the only one—leading to the den.

When summer came, and the undergrowth renewed its foliage, and the grass and the corn grew so tall and thick that Vulp could roam unseen through the fields, he left his haunts amid the woodlands [Pg 234] at the first peep of dawn, and as long as daylight lasted lay quiet in a snug retreat amid the gorse. There all was silent; no patter of summer rain from leaves far overhead, no rustle of summer wind through laden boughs, prevented him hearing the approach of a soft-footed enemy; no harsh, mocking cry of jay or magpie, bent on betraying his whereabouts, gave him cause for uneasiness and fear. Of all wild creatures in the fields and woods, he detested most the meddlesome jay and magpie. If he but ventured by day to cross an open spot, one of these birds would surely detect and follow him, hopping from branch to branch, or swooping with ungainly flight almost on his head, meanwhile hurling at him a thousand abuses. Unless he quickly regained his refuge in the gorse, the blackbirds and the thrushes would join in the tantalising mockery, till it seemed that the whole countryside was aroused by the cry of "Fox! fox!" After such an adventure, it needed the quiet and solitude of night to restore his peace of mind; and even when he had escaped the din, and lay [Pg 235] in his couch among the bleached grass and withered leaves, his ears were continually strained in every direction to catch the least sound of dog or man. When in the winter he ran for life before the hounds, and tried by every artifice to baffle his pursuers, these "clap-cats" of the woods would jeer him on his way. Once, when he ventured into the river, and headed down-stream, thinking that the current would bear his scent below the point where he would land on the opposite bank, the magpie's clatter caused him the utmost fear that his ruse might not succeed. But luckily the hounds and the huntsman were far away. The birds, however, were not the only advertisers of his presence; the squirrel, directly she caught sight of him, would hurry from her seat aloft in fir or beach, to the lowest bough, and thence—though more wary of Vulp than of Brighteye, the water-vole—fling at him the choicest assortment of names her varied vocabulary could supply. Still, for all this irritating abuse Vulp had only himself and his ancestry to blame. The fox loved—as an [Pg 236] article of diet—a plump young fledgling that had fallen from its nest, or a tasty squirrel, with flesh daintily flavoured by many a feast of nuts, or beech-mast, or eggs. It was but natural that his sins, and those of his forefathers, should be accounted to him for punishment, and that it should become the custom, in season and out of season, when he was known to be about, for all the woodland folk to hiss and scream, and expostulate and threaten, and to compel his return to hiding with the least possible delay. Thus it happened that he scarcely ventured, during the day, to attack even a young rabbit that frisked near his lair, lest, screaming to its dam for help, it should bring a very bedlam about his ears.

While roaming abroad in the summer night, Vulp gradually became acquainted with all sorts of vermin-traps used by the keepers. Once, treading on a soft spot near a rabbit "creep," he suddenly felt a slight movement beneath his feet. Springing back, he almost managed to clear the trap; but the sharp steel teeth caught him by a single claw and for a moment held [Pg 237] him fast. He wrenched himself loose, and retired for a while to examine his damaged toe-nail. Then, reassured, he again approached the trap, so that he might store up in memory the circumstances of his near escape. He learned his lesson thoroughly, and never afterwards did the smell of iron, or the slightest taint of the trapper's hand, escape him. He even walked around molehills; they reminded him too much of the soft soil about the trap. And, for the same reason, he avoided treading on freshly excavated earth before the holes of a rabbit warren.

The succeeding years of Vulp's eventful life were in many respects similar to the year that began with his courtship of the sleek young vixen in the white wilderness of the winter fields. His fear of men and hounds increased, while his cunning became greater with every passing day. He never slept on a straight trail, but cast about, returned on the line of his scent, and leaped aside, before retiring to sleep in his retreat amid the bracken. Often he heard the wild, ominous cry of the huntsman, "Eloa-in-hoick, hoick—hoick, cover—hoick!" as the hounds [Pg 238] dashed into the furze; and the loud "Tally-ho!" as he himself, or, perchance, a less fortunate neighbour, broke into sight before the loud-tongued pack. And more than once, from a safe distance, he heard the awful "Whoop!" that proclaimed the death of one of his kindred.

As the years wore on, Vulp gradually wandered far from his old home. The countryside, for twenty or thirty miles around, was known as intimately to him as a little garden, nestling between sunny fruit-tree walls, is known to the cottager who makes it the object of his daily care. His ears were torn by thorns and fighting; his russet coat was streaked with grey along the spine. At last, when age demanded ease and comparative safety from the long, hard chase over hill and dale, he retired to a rocky fastness on the wild west coast, and there, far above the leaping waves and dashing spray, lived his free, lonely life. And there he died.

"HE RETIRED TO A ROCKY FASTNESS ON THE WILD WEST COAST."
To List

It was a bright, hot day in July. Lying among the boulders on the shore, I watched through a field-glass the antics of some birds that wheeled and soared above the cliffs, [Pg 239] when, to my surprise, I saw Vulp crawl slowly along a shelf of rock above a deep, dark cavern. His movements, somehow, appeared unnatural. Instead of crouching, with legs bent under him and brush curled gracefully about his "pads," to bask, his eyelids half-closed, in the sun, he lay on his side. Guided by a companion, who, with waving hand, directed my course as I climbed, I gradually mounted the steep ascent, and peeped over the edge of the rock on which the fox lay. Despite my excessive caution, he was aware of my presence. Slowly and drowsily he lifted his head, uttered a feeble half-grunt, half-whine of alarm, and for a moment bared his teeth defiantly. I remained absolutely still. Then his head fell back, and with a tremor of pain he stretched a stiffened limb. I crawled across the ledge to a rugged path among the cliffs, and descended to the shore. Next day I found him on the rock again, lying in the same position, but dead, while far up in the blue the sea-birds circled and called, and far below, at the edge of the flowing tide, the crested billows leaped and sang.

His "mask" hangs above my study door. [Pg 240] It has been placed there—not as a thing of beauty. The hard, set pose devoid of grace, the bent, dried ears once ever on the alert, the glassy, artificial eyes in sockets once tenanted by living balls of fire that glowed in the darkness of the night—all are unreal and expressionless. Yet the "mask" suggests a hundred pictures, and when I turn aside and forget for a moment the unreality of this poor image of death, I wander, led by fancy, among the moonlit woods, where the red mouse rustles past, and the mournful cry of the brown owl floats through the beeches' shadowed aisles. Then I hear a sudden wail, that echoes from hillside to hillside, as the vixen calls to Vulp: "The night is white; man is asleep; I hunt alone!" And the fox, standing at the edge of the clearing, sends back his sharp, glad answer, "I come!"

THE BROWN HARE.

[Pg 241]

I.

[Pg 243]

THE UPLAND CORNFIELD.

In midsummer, when the sun rises over the hillside opposite my home its first bright beams glance between the branches of a giant oak in the hedgerow of a cornfield above the wooded slope, and sparkle on my study window. And when at evening the valley is deeply shadowed, the light seems to linger in benediction on the same cornfield, where the great oak-tree, no longer silhouetted darkly against a golden dawn, shines faintly, with a radiance borrowed from the west, against the pearl-blue curtain of the waning day. Except during the early morning or at dusk, the cornfield does not stand out conspicuously in the landscape. The eye is attracted by the striking picture of the woodland wall stretching across the slope from [Pg 244] the brink of the river, or by the lower prospect of peaceful meadows and orchards through which the murmuring stream wanders towards the village bridge; but the peaceful uplands beyond rarely greet the vision. For many years I was wont to look from my window only at the woods and the meadows, and somehow I was accustomed to imagine that the line of my vision was bounded by the top of the wood. It was not till more than usual interest had been awakened in me concerning the wild life inhabiting the cornfield, that my eyes were daily turned in the direction of the uplands, where, every evening, the rooks disappear from sight on their way to the tall elms in a neighbouring valley.

Except during harvest, the cornfield is seldom visited by the country folk. It lies away from the main road, and the nearest approach to it is by a grass-grown lane leading from some ruined cottages to a farmstead in the middle of the estate. Many years ago, it was a wilderness of furze and briar, one of the thickest coverts on the countryside, affording safe sanctuary for fox and badger. But gradually it has been reclaimed, [Pg 245] till now only a belt of undergrowth, scarcely twenty yards wide, stretches along the horizon between the upper hedgerow and the wheat.

Here, one starry April night, in a snug "form" prepared by the mother hare, a leveret was born. The "form" was hardly more than a depression in the rank grass, to which, for some time past, the doe had been in the habit of resorting at dawn, that she might hide secure through the day, till the dusk brought with it renewed confidence, and tempted her away into the open meadows beyond the cornfield, where the young clover grew green and succulent. A thick gorse-bush, decked with a wealth of yellow bloom, grew by the side of the "form," and, all around, the matted grass and brambles made a labyrinth, pathless, save for the winding "run" by which the hare approached or left her home.

Unlike the offspring of the rabbit—born blind and naked in an underground nest lined with its parent's fur—the leveret was covered with down, and her eyes were open, from the hour of birth. Nature had fitted her for an existence in the open air. At first she was suckled by day as well as by [Pg 246] night, but as she grew older she seldom felt the want of food till dark. While light remained, she squatted motionless by her mother's side in the "form," protected by the resemblance in colour between her coat and the surrounding herbage, where the browns and greys of last autumn might still be seen among the brambles, with here and there a weather-worn stone or the fresh castings from a field-vole's burrow. In the gloaming, she followed her mother through the "creeps" amid the furze-brake, and sometimes to the edge of the thicket as far as the gap, where she learned to nibble the tastiest leaves in the grass. But soon after nightfall, she was generally alone for some hours while the doe wandered in search of food.

Before daybreak, the doe always returned to suckle her little one. Often in the quiet night, the leveret, feeling lonely or afraid, would call in a low, tremulous voice for help. If the doe was within hearing she immediately responded; but frequently the cry, "leek, leek," did not reach the roaming hare, and the leveret, crouching in the undergrowth, had to wait till she heard her mother's welcome call. Soon the little home in the [Pg 247] thicket was deserted, and the leveret accompanied her mother on her nightly journeys till the fields and the woods for miles around became familiar.

About a month after her birth, the leveret, having grown so rapidly that she was able to take care of herself, parted from her mother, and, crossing the boundary hedge of the estate, took up her quarters on the opposite side of the valley. The doe and her leveret had lived happily in the cornfield and the meadows above the wood. The mother had attended with utmost solicitude to the wants of her offspring, allowing no intruder among her kindred to trespass on her own particular haunts, and careful to select for each day's hiding place some sequestered spot where a human footstep was seldom heard, and the noise of the farmyard sounded faint and remote.

The leveret had learned, partly through a wonderful instinct and partly through her mother's teaching, how to act when there was cause for alarm. Immediately on detecting the presence of an intruder, she lay as still as the stone beside the ant-heap near, trusting that she would not be distinguished [Pg 248] from her surroundings. But if flight was absolutely necessary, she sped away towards the nearest gap, and thence over pasture and cornfield, always uphill if possible, out-distancing any probable pursuer by the marvellous power of her long hind-limbs.

During the late summer and the early autumn, nothing occurred to endanger the leveret's life. The corn grew tall and slowly ripened. Amid its cool shadows the leveret dwelt in solitude. Her "creeps" were out of sight beneath the arching stalks. A gutter for winter drainage, dry and overgrown with grass, formed a tunnel in the hedge-bank between the corn and the root-crop field beyond; and through this gutter the leveret, when at night she grew hungry, could steal into the dense tangle of thistles and nettles fringing the turnips, thence, between the ridges under the wide-spreading leaves, to the narrow pathway dividing the rape from the root-crop, and across the field to a furrow where sweet red carrots, topped with dew-sprinkled plumes, tempted her dainty appetite.

When the calm night was illumined, but not too brightly, by the moon and stars, [Pg 249] the leveret would venture far away from her retreat to visit a cottage garden where the young lettuces were crisp and tender. Her depredations among the carrots and lettuces were scarcely such as to deserve punishment. She ate only enough of the lettuces to make a slight difference in the number of seeding plants ultimately devoured by the cottager's pig, or thrown to the refuse-heap; and from the great pile of carrots, to be gathered and stored in the peat-mound by the farmstead, the few she destroyed could never by any chance be missed. On all the countryside she was the most inoffensive creature—the harmless gipsy of the animal world, having no fixed abode, her tent-roof being the dome of the sky.

As autumn advanced, the reapers came to the corn. She heard them enter by the gate; and presently, along the broad path cut by the scythe around the field, the great machine clanked and whirred. All day the strange, disturbing noise continued, drawing gradually nearer the spot where the leveret lay. Through the spaces between the stalks she watched the whirling arms swinging over, and the horses plodding leisurely by the edge [Pg 250] of the standing wheat. At last, but almost too late, she leaped from her "form" as the cruel teeth cut through the stalks at her side; and, taking the direction of her "creep," rushed off towards the nearest gap and disappeared over the brow of the hill.

In the middle of the night she wandered back to the wheat-field. The scene before her eyes revealed a startling change. The corn stood in "stooks" on the stubble; no winding paths led here and there through a silent sanctuary, where countless waving, nodding plumes, bent and released by a gentle-flowing wind, had shimmered in the bright radiance of the harvest moon, when, coming home late at night from the marsh across the hill, she had stayed for a while on the mound by the gate, and tiptoe, with black-fringed ears moving restlessly, had listened to some ominous sound in the farmyard. The prickly stubble felt strange to her feet, so, carefully picking her way by the ditch, she crossed to the nearest gate and ambled down the lane. But the change noticed in the wheat-field seemed to have passed over the whole countryside. It was more and more pronounced during the [Pg 251] following week, till, in October, the late harvest had all been cleared. The habits of the hare altered with the season. Having at last grown accustomed to the varied conditions of her life, she sometimes frequented the old tracks over the upland, but rarely resorted to the "forms" in which she had lain amid the summer wheat.

October brought her an experience which might have proved disastrous, but which, fortunately, resulted in nothing more than a passing fright. In the stalk of the rye occurs a knot, forming a slight bulge known to the peasantry as the "sweet joint." Rabbits and hares are extremely fond of this succulent morsel, and, in consequence, the rye-crop, if near a large warren, is in danger of being totally destroyed. Puss one night had wandered far to a field, where, some time before, she had discovered a patch of standing rye. The few remaining stalks were hard and uninviting, but there were some delicious parsnips among the root-crops. At dawn she settled down to hide between the rows of swedes close by, and remained secreted for the day; but towards evening a sportsman came in at the gate, and, with a low word of command [Pg 252] and a wave of the arm, "threw off" his brace of red setters to range the field. Working systematically to right and left, the dogs sought eagerly for game. Soon the hare was scented, and while Juno, with stiffened "stern" and uplifted paw, stood almost over her, Random, "backing" his companion, set towards the furrow where Puss, perfectly rigid, and with ears well over her shoulders, crouched low, prepared for instant flight. Step by step the sportsman, with gun in readiness, moved towards Juno, cautioning her against excitement; while Random, sinking on his haunches, awaited patiently the issue of events. Suddenly, convinced that in flight lay her surest chance of escape, the hare leaped from her "seat," and with the utmost speed, though from the ease of her motions appearing to run slowly, made her way towards the hedgerow. There was a quick rush behind her as she started from the furrow, and then a loud, rasping exclamation from the sportsman, but nothing more; no shot was fired. She owed her life to several circumstances. The dogs were young, and in strict training; their master, knowing the natural fondness [Pg 253] of "first season" setters for "chasing fur," had purposely refrained from killing the hare, and had turned his attention to the behaviour of his dogs. Then, again, he cherished a certain fondness for Puss, believing her to be the most persecuted, as well as the most innocent and interesting, of Nature's wildlings in the wind-swept upland fields.

Henceforward, but for one other incident, the life of the hare was singularly uneventful till the early spring. That incident occurred within a week of her escape from the setters, and once more her luck was due to the humanity of him who had found her among the turnips. The farm-lands frequented by the leveret were a favourite resort of many of her kind, and when moving about in the darkness of the night she often found signs of their presence near the gaps and gateways. The sportsman, knowing well that after harvest the poaching instincts of the peasantry and of the professional village "mouchers" would receive fresh stimulus, determined to forestall his enemies, and render futile some, at least, of their endeavours. So it came about that one night a keeper, assisted by several of the guests at the "big house" in the valley, and having previously made every [Pg 254] preparation for the event, placed a net near each gate and before each likely gap within a radius of half a mile from the heart of the estate.

Unless hard pressed, a hare seldom leaves a field except by certain well-known openings in the hedgerow. Unlike the rabbit, she will not readily leap over any obstacle beneath which she can crawl; and whereas the "creep" of a rabbit through a gateway or a hedgerow is well-nigh invariably at right angles to the line of that gateway or hedgerow, the "creep" of a hare tends sideways and is sometimes slightly curved. To net hares successfully it is necessary to know their habits; and the keeper, having served a lifelong apprenticeship in field-craft, was prepared for every emergency. His object at this time was not to kill the hares, but simply to educate them, to warn them thoroughly once for all against the wiles of their worst enemy, the poacher.

As Puss was busily feeding in the dewy clover, she heard the quick, continuous gallop of a dog. This time, however, she had not to deal with Juno, the setter, but with a trained lurcher, borrowed for the occasion from a keeper who had captured the animal [Pg 255] during a poaching affray. The leveret, peeping over the grass-tops, saw the dog coming rapidly on. He was over and past her in an instant. As he turned, she started off straight towards an opening where some sheep had partly broken down the hedge. The lurcher closed in, and drove her thither at tremendous speed. She strained every nerve, and, gaining the ditch, blundered blindly through the gap, and fell, helpless and inert, entangled completely within the treacherous folds of the unseen net. Her piteous cries, tremulous, wailing, heart-rending—similar to the cries of a suffering infant—were borne far and wide on the wind. The keeper soon reached the spot, and, placing his hand over her mouth to stop the cries, tenderly extricated the frightened creature from the treacherous meshes and allowed her to go free. For a few seconds, she lay in abject fright, panting and unable to move. Then, hearing the cries of another hare entangled in a bag-net some distance away, she bounded to her feet, and darted off—somewhere, anywhere, so long as she might leave the awful peril behind. Bewildered, but with every instinct assisting [Pg 256] her in the desire for life, she ran along by the hedgerow, and, unexpectedly catching sight of a familiar gate, crouched and passed quickly through the "creep" beneath the lowest bar. But here, again, a net was spread; again the hare fell screaming and struggling into the meshes; and again the keeper released her. Exhausted by intense excitement and fear, she crawled into the "trash" in the ditch, and kept in hiding, not daring the risk of another capture. Luckily for Puss, the lurcher had already hunted the field in which she was now secreted, and so the timid creature remained undisturbed beneath the fern. When her wildly throbbing heart had been quieted by rest and solitude, she stole from her hiding place to nibble the clover at the side of the path. Towards dawn, she journeyed to a wide stretch of moorland on the opposite hills, and there made a new "form" on a rough bank that separated a reedy hollow from the undulating wilderness of heather and fern.

The leveret's adventures were destined to effect a considerable change in her habits. She was being roughly taught that to preserve [Pg 257] her life she must be ever cautious and vigilant. Though danger threatened her by day and by night, she lived beyond the usual period of a hare's existence, partly because her early education was thorough and severe. Thus taught, she would pause for an instant at every gap and gateway before she passed through, and, if she found a net in her path, would turn aside, creep along by the hedge, and seek an exit at another place.

The perils to which she had been exposed created a feeling of intense restlessness, which harassed her throughout the winter months, and caused her to travel long distances, by the loneliest lanes and fields, to and from the moorland where now she had made her home. She remembered the scent of a human being since her experiences with the keeper, and, her powers of smell being wonderfully acute, was able to detect even the faintest signs which indicated that her dread enemy—man—had crossed her path. One night she smelt the touch of a hand on the grass-bents near her "form," and found also that the herbage had been moved aside. Though the scent was faint—the intruder having visited the spot soon after the leveret had [Pg 258] set out in quest of food—the cautious creature forsook her lair, and spent the day in a sheltered retreat beside a heap of dry and withered leaves near the outskirts of a copse on the slope overlooking the moor.

Gradually she grew big and strong, becoming unusually fat as the autumn advanced, so that she would be able, if required, to withstand the rigour and the waste of a severe winter. Her coat was thick and beautifully soft, for protection against cold and damp. But while she increased in weight, she remained in hard condition because of her long journeys and frequent change of quarters.

It happened, however, that her first winter was helpful to the welfare of animal life in general. The heavy rains, it is true, greatly distressed the leveret. The nights were so dark, and the constant patter of the rain so interfered with even her highly trained powers of hearing, that, while the wet weather lasted, she seldom dared to leave the neighbourhood of her favourite resort, but crouched in the grass at the margin of the copse, and tried to obtain a meal as best she could from the sodden herbage.

Though on certain occasions Puss might [Pg 259] have been discovered in hiding on the marsh, yet there, whenever possible, she chose a dry spot for her "seat." She loved, best of all, the undulating hills far above the river-mists, which, chilled at nightfall by an occasional frost, descended on the fields like crystal dust, and almost choked her if she chanced to pass within these wreathing drifts that brought discomfort and disease to man and beast alike.

But the want of exercise so affected her, that, when again the weather was fine and she ventured from her lair, she found herself unable to cover the usual distance of her nightly rambles. As the first cold glimmer of the dawn appeared in the south-eastern sky, she started back, in alarm at her fatigue, to complete the remaining mile of her journey home. Her weakness soon became apparent.

Then, finding herself powerless to proceed, she turned reluctantly aside, and crouched, with Nature's mimicry for her protection, on the brown ploughland where the winter wheat was thrusting up its first green sprouts above the soil. But after a few days she was well and strong again. She [Pg 260] suffered far less from the short, sharp frost that bound the countryside with its icy fetters, than from the rains. The frost scarcely interfered with her movements; indeed, it made exercise more than ever necessary. Forced to seek diligently for her food, she found it in a deserted stubble; there, when the sheep lay sleeping in the bright winter moonlight, she would squat beside them, nibbling the turnips scattered over the field as provender for the flock.

II.

MARCH MADNESS.

March came in "like a lion." The wind whistled round the farmstead on the hill, and through the doorway of the great kitchen, and down the open chimney. It woke up the old, grey-haired farmer who dozed on the "skew" in the ingle-nook by the crackling wood-fire; it almost made him feel young again with the vigour of the boisterous spring. It sang in the key-hole of the door between the passage and the best parlour; the mat at the threshold flapped with a sound as of pattering feet; and the gaudy calendars on the wall flew up like banners streaming in the breeze. The old man turned, and eagerly watched the hailstones, as they dropped tinkling on the roofs of the outhouses, or, driven aslant by the wind, crashed hissing against the ground, and, rebounding, rolled across the pebbled yard. The labourers came home to the mid-day meal, and, pausing at the door, shook the hail from their garments.

"Lads," said the farmer, "I've been spared to hear the whisper of another spring."

"God be thanked!" said the hind, "for seasonable weather at last. Every man to his trencher! the broth is in the bowls."

Out on the marsh the reeds beat in the wind. Every grass-fibre twisted and swung; the matted tussocks, drooping over stagnant pools near which the snipe, with ruffled feathers, probed the soil in search of food, were shaken and disentangled, so that the bleached blades of last year's growth fell apart, and exposed the fresh young sprouts rising from the bed of winter's death. Over the wide waste the March wind drove furiously, with blessing in the guise of chastisement, while, far above, the grey-blue clouds whirled fast across a steely sky, till the ashen moon gazed coldly on the waning day, as one by one the stars flashed overhead, the clouds rolled down into the pink and silver west, and the song of the wind became only a murmur in the leafless willows by the brook.

With the advent of March, a great change passed over the wild life of the uplands. The jack-hares threw aside their timidity, and wandered, reckless of danger, over the marsh, across the stubbles, and through the woods. Even in broad daylight, they frisked and quarrelled, in courtship and rivalry.

The leveret was now full-grown, and Nature's mothering instincts were strong within her. One evening, as she louped along her accustomed trail towards the turnip-field, she discovered a suitor following in her wake. Half in misgiving, half in wantonness, she turned aside and hid in the ditch. Presently she felt a soft touch on her neck: the jack-hare was pushing his way through the undergrowth. For a moment she stopped to admire him as the moonlight gleamed on a white star in the centre of his forehead. Then away she jumped, dodging round the bushes and hither and thither among the grassy tangles, while her admirer followed, frisking and leaping in sportive gaiety. Another jack-hare now came along the hedgerow. In utter mischief, [Pg 264] Puss called "leek, leek, leek," as if pretending to be in distress and in need of help. "Leek, leek," came the low response, as, quickening his pace, the second hare sprang into the fern. But his audacity was not to go unchallenged. The first suitor immediately showed himself, and, making a great pretence of reckless bravery, prepared to give the second a warm reception. The doe-leveret, apparently indifferent, but nevertheless keenly interested in the combat, crouched on a little knoll by the path, while the jack-hares, sitting on their haunches, boxed and scratched, and rolled over each other in a singularly harmless conflict, neither suffering more than the loss of a few tufts of fur. The comedy might, however, have had a tragic ending. Presently one of the combatants—the hare that had come late on the scene—became slightly exhausted, and, ignominiously yielding to his rival's superior dexterity, ran back towards the distant hedge. Almost at once a fox crept out from the furze at the corner of the field, and trotted away on the scent of the fleeing hare, while Puss and her mate made off in the direction of a more secluded pasture. [Pg 265]

A month passed—a month of general hilarity and indiscriminate fighting among all the hares in the district—and then, within a neat, dry "form," that Puss, with a mother's solicitude, had made in a carefully selected spot on a mound where the grass was tall and thick, her little leveret was "kittled." The doe-hare tended her offspring as carefully as she herself had been tended a year before. Her faithless lover had gone his own way. But Puss cared little for his desertion: she wished to live alone, under no monopoly as far as her affections were concerned, though for the time her leveret wholly engaged her mothering love.

So strong was her strange new passion that she was ready, if needs be, to brave death in defence of her young. And, not long after the leveret's birth, the mother's courage was tested to the utmost. A peregrine falcon, from the wild, rocky coast to the west, came sailing on wide-reaching wings across the April sky. Puss was resting in a clump of brambles not far from her "form," and saw the big hawk flying swiftly above. Any movement on her part would have instantly attracted the attention of her foe, so she [Pg 266] squatted motionless, while her leveret also instinctively lay still in its "form." But the keen eyes of the falcon detected the young hare, and the bird descended like a stone on his helpless victim. Instantly, the doe rushed to the rescue, and, effectually warding the attack, received the full force of the "stoop" on her shoulders. As the hawk rose into the air, the doe felt a sharp pain in one of her ears—the big talons, closing in their grasp, had ripped it as with the edge of a knife. She screamed, then, grunting savagely, leaped hither and thither around the leveret, meanwhile urging it to escape into the adjacent thicket. The bird, aloft in the air, seemed perplexed, and eventually prepared to "stoop" again. In the nick of time, Puss vanished with her little one beneath an impenetrable tangle of friendly thorns, while the baffled peregrine proceeded on his way.

For some weeks, the hare languished under the effects of the falcon's blow. When her leveret was old enough to find food for itself, she rested, forced by the wound to live quietly in hiding, till the scar healed and [Pg 267] life once more became enjoyable. But she always bore the marks of the talons, and so was spoken of by the country folk as "the slit-eared hare."

The superstitious recalled the tales of a bygone century, and half believed the hare to be a witch in disguise, for she seemed to bear a charmed life, and, though known everywhere in the parish, successfully eluded to the end all the devices that threatened her. No matter how artfully the wire noose was set above the level of the ground in her "run," she brushed it by and never blundered into the treacherous loop. A net failed even to alarm her: it might almost be imagined that she became an experienced judge of any such contrivance, and knew every individual poacher by the method with which his toils were spread across her path.

Not having bred during the year in which she was born, Puss had thrived, and weighed about nine pounds in the late autumn of her second season. But according to popular opinion she was much heavier. Will, the cobbler, who was fond of coursing, stoutly maintained, to a group of interested listeners in the bar-parlour of the village inn, that she [Pg 268] seemed like a donkey when she escaped from his greyhound into the wood.

Family cares again claimed the hare's attention in July; and, having taken to heart her experience with the peregrine, she left the uplands and made her home in the thickets of a river-island. At that time the river was low, and, on one side of the island, the bed of the stream had become a dry, pebbly hollow, save for a large pool fed by the backwater at the lower end, where the minnows played, and whither the big trout wandered from the rapids to feed during the hot summer nights.

Late one afternoon, when long shadows lay across the mossy bank of the river beyond the tall beeches standing at the entrance to the island thickets, Puss was waiting for the dusk, and dozing meanwhile, but with wide-open eyes, beside her leveret. Since there was another little mouth besides her own requiring food, she generally felt hungry long before nightfall, and so, when the afterglow began to fade in the west, was wont to steal away to the clover above the woods that fringed the long, still pool up-stream.

As the day wore on, the hare heard the [Pg 269] unmistakable tread of human feet approaching through the woods. The sounds became increasingly distinct; then a pebble rattled and splashed into the water as the intruder walked across the river-bed. He passed close to the "form," and, turning down-stream, was lost to sight amid the bushes. At intervals, the hare imagined that the faint, muffled sounds of footsteps came from the distance; but again the sounds drew near, ceasing, however, when the man was a few yards from the nest.

I can complete the story. Since spring I had been studying the wild life of this lonely island below the rocky gorge extending hither from the village bridge. The wood-wren, the willow-wren, and the garden-warbler had nested in the thickets, and every evening I had visited the place to pry on their doings, and to note how the flowers in glad succession blossomed and faded—their presence in this lonely sanctuary known only to myself, and to the birds, bees, and butterflies, and to the little shrews that rustled over the dry leaves beneath. But now the garden-warblers had left for the copse on the far side of the river, and the wood-wrens [Pg 270] and the willow-wrens had retreated to the inner recesses of the thickets, where, amid the luxuriant verdure of midsummer, their movements baffled my observation.

On the July evening, as I lay in the matted grass at the edge of the copse by the pebbles, watching a whitethroat among the bushes opposite, my eye happened to rest for an instant on a patch of bare mud immediately before me. There, to my surprise, I discovered the footprints of the hare. The five toes of the fore-feet, and the four toes of the hind-feet, were as clearly outlined as if each impression had been taken in plaster. And yet, when I stood up to look at the spot, the marks seemed to have wholly disappeared. On nearer examination I found that the track of the hare was in the direction of the island. From their shape, and the distance between each, the footprints indicated that the movements of the hare had not been hurried. Similar footprints were visible in a straight line between the bank and the island. Only one conclusion seemed possible—the hare had crossed to the island early that morning, after the heavy shower that had fallen just before dawn. It would have been [Pg 271] contrary to her habits had she crossed later; and, had she passed the place at any time before, the rain would have washed away the marks in such an open spot, or, at any rate, would have blurred them beyond recognition.

After placing a white stone by the footprints to indicate their whereabouts, I searched along the river-bed for signs that would show a track towards the bank; but not a single mark could be found pointing in that direction. It was obvious that the hare had not left the island till, at any rate, some hours after the rain. Then, however, the sun would have been so high that Puss would have been loath to leave her lair. Faintly discernible beside a large pebble, one other footprint appeared, leading like the rest towards the island. The mark was old, and had been saved from obliteration by the sheltering stone; but it suggested that the hare had made her home not far away. Taught by experience, I decided not to penetrate the copse and risk disturbing its probable tenant. I approached it only so far as to examine another bare place in a line with the footprints on the mud, where, [Pg 272] to my delight, I found fresh footprints similar to those at the dried-up ford, together with other and much smaller marks undoubtedly made by a tiny leveret.

I now re-crossed the ford and went home. But before nightfall I returned, and, hiding behind the hedgerow on the bank, watched, unseen, the approach to the island. My patience was soon rewarded. Just as the dusk was deepening over the woodlands, "the slit-eared hare" left her "form" and stood in full view by the ford. There, having lazily stretched her long, supple limbs, she played awhile with her leveret, sometimes pausing to nibble a few clover-leaves as if to direct the little one's attention towards its suitable food. Then she ambled leisurely across the river-bed, and, with graceful, swinging gait, passed through the meadow beyond—while her offspring disappeared within the thickets of the island.

The hot weather broke up in July, and henceforth, till late September, rain descended almost every day. The shower that had revealed the whereabouts of the hare was the first sign of the change. On the following night, a thunderstorm broke over the countryside, [Pg 273] washed down the soil from the pastures, and sent the river roaring in flood through the gorge. While on the far side of the island the main torrent raged past beneath the willows, the divided stream under the near bank formed salmon-pools and trout-reaches, where, before, the pebbles had been bare and dry.

Anxious to know how the flood would interfere with the movements of the hare, I came back on the following evening to my hiding place by the hedgerow. In the dusk, Puss appeared at the margin of the copse, and moved down the bank to the edge of the stream. There she paused, apparently perplexed, and called to her leveret. Presently the young hare joined her mother at the water's edge, and both hopped along the brink, seeking a dry place by which they might reach the field on the slope. Finding none, they adjourned to the mossy bank where I had seen the leveret's footprints. Then the doe went down boldly to the stream, called to her companion, waded in, and swam across. Ascending into the field, she shook the water from her fur, and again called repeatedly. The young [Pg 274] one hesitated, and ran to and fro crying piteously, "leek—leek." Suddenly, in the excitement, it missed its footfall and fell into the river. Bewildered, but hearing its mother's call, it swam down the pool through the still water below the little rapid, and landed on the opposite bank, where it joined its parent, and, following her example, shook the water from its downy limbs. Soon both disappeared within the wood; and, satisfied with my evening's sport, I turned homewards across the fields.

During the rest of the summer, the hare frequented the rough pastures skirting the ploughlands, and visited the cornfields only when the weather was dry. Hares suffer little discomfort in rainy weather, if only the fine fur beneath the surface of the coat remains dry—after a shower they can easily shake off any outside moisture. But they dislike entering damp places where the vegetation is tall and their fur may get matted and soaked by the raindrops collected on the herbage. In wet weather hares may often be found in cover, especially near thick furze-brakes on a well drained hillside, but [Pg 275] their presence in such a situation may imply that they sought shelter before the rain began to fall.

In September, for the third time during the year, Puss was occupied with family affairs. Now, three tiny leverets were "kittled," and the nest occupied an almost bare place on the top of a ridge in the root-field where last season the succulent carrots grew. The hare had been greatly distressed by the unusually wet summer, and one of her leverets was in consequence a weakling; another leveret was killed by a prowling polecat while the mother wandered from the "form"; and only the third grew up robust and strong.

The approach of winter brought Puss many strange experiences, from some of which she barely emerged with her life. When the season was passed, it had become more than ever difficult to approach her; she would slip away to cover directly her keen senses detected the presence of a stranger in the field where she lay in her "form." As she grew older, her leverets sometimes numbered four or five, but as a rule she [Pg 276] gave birth to three only, her productiveness being probably dependent on the ease with which she obtained food.

One day in February, just before bringing an early little family into the world, she almost met her death. A village poacher, ferreting on the hillside, chanced to see her, as she lay not far off in a patch of clover. Without waste of time, he proceeded to attempt the capture of the hare by a well-known trick. Thrusting a stake into the ground, he placed his hat on it, and strolled unconcernedly away. Then, as though he had changed his mind, he walked round the clump, in ever narrowing circles, gradually closing on his prey. Meanwhile, the hare, her attention wholly diverted by the improvised scarecrow, remained motionless, baffled by the artifice. Suddenly she felt the touch of the man's hand. The poacher had thrown himself down on the tuft, hoping to clutch the hare before she could move. But in endeavouring to look away from the spot, and, at the same time, measure the distance of his fall, he had miscalculated the hare's position. She sprang up, and with ears held low sped away towards the wood, [Pg 277] leaving the poacher wild with rage at the failure of his ruse, and vowing vengeance on the timid creature, whose life, at such a time, would hardly, even to him, have been worth an effort.

III.

THE CHASE.

Of all the hounds employed in the chase of the hare, the basset promises to become the prime favourite among some true-hearted sportsmen who love sport for its own sake, and not from a desire to kill. He is a loose, lumbering little fellow—resembling his relative, the dachshund—low and long, with out-turned legs, sickle-shaped "flag," and features which, in repose, seem to suggest that he has borne the grief and the care of a hundred years, but which, when the huntsman comes to open the kennel doors, are radiant with delight. Mirthfulness and dignity seem to seek expression in every movement of the quaint, old-fashioned little hound, and in every line of his face. As for his music—who would expect such a deep, bell-like note from this queer midget among hunters, standing not much higher than the second button of the huntsman's legging? Withal, he is a merry, lively little fellow, with a good nose for the scent of a rabbit or a hare, and, when in fit condition, is able to follow, follow, follow, if needed, from earliest dawn till the coming of night. The chase being ended, he with his companions, Harlequin and Columbine, and all the stragglers of the panting pack, will surround the tired hare, and will wait, bellowing lustily, but without molesting the quarry, till the Master appears and calls them to heel.

If the ten to twenty sportsmen often to be found in a village would combine, each keeping a basset for the common Hunt, they might derive the utmost pleasure from following their pets afield, and incidentally would assist to prevent the extermination of an innocent wildling of our fields and woodlands. For the sake of the sport shown by the basset-hounds, many of the farmers near the villages, who dearly love to hear the deep music of a pack in full cry, would protect Puss from those more cunning and powerful enemies of hers, who, lurcher in leash or gun in hand, steal along the hedgerows at nightfall, so that, from a secret transaction thereafter with some local game-dealer, they may get the wherewithal for a carouse in the kitchen of the "Blossom" or the "Bunch of Grapes."

One morning in December, when the rime lay thick on the fields, and the unclouded sun, rising in the steel-blue sky, cast a radiance over the glittering countryside, our village basset-hounds found the "cold" scent of the hare in the woods above the church, where Puss had sheltered beside a prostrate pine-trunk before returning to her "form" at dawn. After endeavouring in vain for some time to discover the direction of her "run," they set off, "checking" occasionally, across the stubble, through the root-crop field, and down over the fallow to the bottom of the dingle. There, near a bubbling spring, Puss had hidden since daybreak. Hearing the far-off music, she slipped out of the field unobserved, till, reaching the uplands, she was seen to pass leisurely by in the direction of the furze-brake.

Directly the bassets came to the spring, a chorus of deep sounds announced that the [Pg 281] quarry had been tracked to her recent lair. All through the morning they continued their quest; they streamed in a long, irregular line up the hillside, their black and tan and white coats gaily conspicuous in the sunlight; they trickled over the hedgerows, and dotted the furrows of the deserted ploughlands; they moved in "open order" through the copse, and plodded along by the furze-brakes or through the undergrowth where the sharp-thorned brambles continually annoyed and impeded them; they worked as if time needed not to be taken into the slightest account. The least scent met with loud and hearty recognition; fancy ran riot with the excited puppies; the atmosphere at every turn seemed to betray the near presence of Puss. But every condition of weather and fortune was against good sport. The ground was steadily thawing in the warmth of the sun, and the rising vapour, trembling in the light, seemed to carry the scent too high for accurate hunting.

So the hare ambled along her line of flight—a wide, horse-shoe curve that began and ended in the fallow on the slope. When a considerable distance had been placed [Pg 282] between herself and her pursuers, she ceased to hurry. Indeed, the music of horn and hounds seemed almost to fascinate the creature, and frequently she lingered for a few moments to listen intently to the clamour of her enemies. A farm labourer, who tried to "grab" her as she passed down the grassy lane, said that she "was coming along as cool as a cucumber. Sometimes she'd sit down to tickle her neck with her hind-feet. Then she'd give a big jump, casual-like, to one side of the path, and sit down again, with her ears twitching and turning as if she thought there was mischief in every flutter of a leaf or creak of a bough."

Frightened almost out of her wits by the labourer's sudden and well-nigh successful endeavour to secure her, Puss rushed back along the lane, crossed a gap, and sped over the uplands once more, leaving her usual horse-shoe line of flight, and taking a much greater curve towards the fallow. But gradually her pace slackened as she discovered she was no longer followed; and then, not far from her lair by the spring, she paused to rest. The music of the hounds was faint, distant, and intermittent; and at [Pg 283] last it entirely ceased. Somewhat exhausted towards the end of her journey, she had withheld her scent, and had thus completely outwitted her slow but patient pursuers.

Once, and once only, towards the end of January, she found herself chased by her more formidable foes, the beagles. At first she eluded them by stealing off without yielding the faintest scent; but she was "viewed" in crossing the meadow, and the hounds, making a long, wide cast, "picked up" as soon as a slight, increasing taint in the air was perceptible, then followed for several miles. But, ultimately, they were baffled, and Puss made good her escape.

It had happened that, after creeping through a gutter in the hedgerow of a stubble, she had come in sight of a flock of sheep grazing on the opposite side. Like Vulp, the fox, she knew how to hinder the chase by mingling her scent with that of other animals; so without hesitation she passed through the flock, and made straight for an open gateway in the far corner of the field. When the beagles, in hot pursuit, appeared on the scene, the startled sheep, rushing away, took the line of the hunted hare through the [Pg 284] opening, and thus "fouled" the scent so thoroughly that the hunt came to a "check." After the hare had left the fields frequented by the sheep, she took the direction of a path leading over a wide bog towards the woodland. On the damp ground the scent lay so badly, that when, some time later, the beagles crossed her line, they were unable, even after repeated "casts," to follow her track. Presently the impatient huntsman, with hounds at heel, moved away to the nearest road and relinquished his quest.

Luckily for Puss, the harriers never visited her neighbourhood, and only on special occasions was coursing permitted on the estate. If at night a lurcher entered the field in which she grazed amid the clover, her knowledge of the poacher's artifices immediately prompted her to slip over the hedge and past the treacherous nets. Her life, beset with hidden dangers, was preserved by a chain of wonderfully favourable circumstances, that befriended her even when the utmost caution and vigilance had been unavailing.

Once, so mild was the winter that the hare's first family for the year came into the [Pg 285] world in January. A few weeks afterwards, when she was about to separate from her leverets, an incident occurred that might have been attended with fatal results. A poacher, prowling along the far side of the hedgerow, and occasionally stopping to peep through the bushes for partridges "jugging" in the grass-field, caught sight of the leverets nibbling the clover near a small blackthorn. In the feeble afterglow, he was uncertain that the objects before him were worth the risk of a shot, so he crawled towards a gap to obtain a nearer view. To his astonishment, when he reached the gap nothing was visible by the thornbush; the leverets had vanished in the ferns. But the poacher was artful and experienced. He hid in the undergrowth of the ditch, where, after waiting awhile, and seeing no sign of movement in the grass, he gave utterance to a shrill cry like that of a young hare in distress. Five minutes passed, and the cry was repeated— tremulous, prolonged, eloquent of helpless suffering. At intervals, the same artifice was employed, but apparently without success.

The poacher was about to crawl from his hiding place, when suddenly, close beside the [Pg 286] hedgerow, the head of the doe hare came into sight. Startled, in spite of expectation, by her sudden appearance, and excited as he recognised the "slit-eared hare," the poacher involuntarily moved to grasp his gun. He looked down for an instant to make sure that his gun was in readiness, but when he lifted his eyes again the hare was gone. Do what he might, not another glimpse of his quarry was to be obtained, and so, half believing that he had seen a witch or that he had dreamed, he stole away into the darkening night.

Deceived by the poacher's cries, the doe-hare had hurried home, but had found her young alive and well. Then, scenting danger, she had vanished with her offspring into the nearest bramble-clump, and in the deep shadow of the hedgerow had led them safely away.

During the last year of her life, she frequented the hawthorn hedges and the furze brakes of an estate diligently "preserved" by a lover of Nature as a sanctuary whither the furred and feathered denizens of the countryside might resort without fear of hounds or poachers, and where a gun was never fired except at vermin. The winter [Pg 287] was severe; on two occasions snow lay thick on the ground for more than a week. But Puss was fairly comfortable; she had her "form" on a dry, rough heap of stones, gathered from the fields and thrown into a disused quarry near the woods; and for four or five nights she remained at home, the snow covering her completely but for a breathing hole in the white walls of her tiny hut. At last, impatient of confinement, and desperately hungry, she broke through the snow-drift, and sought the nearest root-crop field, where, after scratching the snow from a turnip, she was able to make a hearty meal. While returning slowly towards the wood through the soft, yielding snow that rendered her journey difficult and tiresome, she unexpectedly discovered, near the hedge beyond the furrows, a tasty leaf or two of the rest-harrow, together with a few yellow sprouts of young grass where a stone had been kicked aside by a passing sheep—these were the tit-bits of her provender.

In the early morning, the hare, too cautious to re-enter the "form," which, now that its surroundings were torn asunder, had become a conspicuous rent in the white mantle of [Pg 288] the old quarry, crept over the hedge into the woods, and, moving leisurely beneath the snow-laden undergrowth, where her deep footprints could not easily be tracked, selected a suitable spot for a new "form" in the friendly shelter of a fallen pine.

But even in this woodland sanctuary she encountered an enemy. A cat from the farm on the hill, having acquired poaching habits, had strayed, and taken up her abode among the boulders at the foot of a wooded precipice adjoining the lower pastures of the estate. In a gallery between these boulders, she had made her nest of withered grass and oak-leaves, where, at the time of which I write, she was occupied with a family of kittens. The wants of the kittens taxed the mother's utmost powers; she prowled far and wide in search of food, and was as much a creature of the night as were the fox and the polecat that also lived among the rocks.

There is no greater enemy of game than the renegade cat. She is far more destructive than a fox. Many animals that can evade Reynard are helpless in the grip of a foe armed so completely as to seem all fangs [Pg 289] and talons. The special method of slaughter adopted by the cat towards a victim of her own size is cruel and repulsive in the extreme. Grasping it with her fore-claws and holding it with her teeth, she lies on her back and uses her hind-claws with such effect that often her prey is lacerated to death.

Roaming at night in the shadow, the cat came unexpectedly on the scent of the hare and traced it to the "form," but the desired victim was not at home. The cat returned to the spot before dawn, and lurked in hiding beneath the hawthorns. The hare, however, was not to be easily trapped. Coming into the wood against the wind, she fortunately detected the enemy's presence quite as readily as the cat had discovered her "form" amid the grass-bents. With ears set close, and limbs and tail twitching with excitement, the cat crouched ready for the deadly leap. But the hare suddenly sprang aside from her path, climbed the hedgerow, and disappeared, outpacing with ease the cat's half-hearted attempt at pursuit.

At length the "slit-eared hare" met her death, in a manner befitting the wild, free existence she had led among the hills and [Pg 290] the valleys. Her dead body was brought me by the head keeper of the woodland estate, and, as it rested on my study table, I gazed at it almost in wonder. The russet coat, turning grey with age, was eloquent of the brown earth, the sere leaf, and the colourless calm of twilight, and told me of the creature's times and seasons. The big, dark eyes, their marvellous beauty and expressiveness dimmed by death, and the long, sensitive ears, one ripped by the falcon's talon and both slightly bent at the tip with age, were suggestive of persecution, and of a haunting fear banished only with the coming of night, when, perchance, the early autumn moon rose over the corn, and the hare played with her leverets among the shadowy "creeps." My hands rested on the fine, white down that took the place of the russet coat where Nature's mimicry was needed not; it was pure and stainless, like the lonely wildling's inoffensive life.

"WHEN THE EARLY AUTUMN MOON ROSE OVER THE CORN." To List

A terrible thunderstorm had raged over the countryside all the evening and throughout the night. Ben, the carter, coming home to the farm with his team, had dropped at the very threshold of the stable, blasted in a [Pg 291] lurid furnace of sudden fire. A labourer's cottage had been wrecked; many a stately forest tree had been rent or blighted; the withering havoc had spread far and wide over the hills. On the following morning, the keeper, going his rounds, had found the dead hare beside a riven oak.

THE BADGER.

[Pg 293]

I.

A WOODLAND SOLITUDE.

Even in our own densely peopled land, there are out of the way districts in which human footsteps are seldom heard and many rare wild creatures flourish unmolested. Near such parts the naturalist delights to dwell, in touch, on one side, with subjects that deserve his patient study, and, on the other side, with kindly country folk, who, perhaps, supply him with food, and are the means of communication between him and the strenuous world. In this western county, however, the naturalist, in order to gain expert knowledge, does not need to live on the fringe of civilisation. Here, among the scattered upland farms around the old village, creatures that would elsewhere be in daily danger because of their supposed attacks on game are almost entirely free from persecution. In several of our woods, polecats seem to be more numerous than stoats, and badgers are known, but only to the persistent observer, to be more common than foxes; and both polecats and badgers are seldom disturbed, though the farmers may regularly pass their burrows.

The immunity of such animals from harm is, to some extent, the result of the farmer's lack of interest in their doings. He strongly resents the presence of too many rabbits on his land, "scratching" the soil, spoiling the hedges, and devouring the young crops, and, therefore, cherishes no grudge against their enemies so long as his stock is unmolested. He is no ardent protector of game, and, if a clutch of eggs disappears from the pheasant's nest he has chanced to discover in the woods, thinks little about the incident, and concludes that Ned the blacksmith's broody hen has probably been requisitioned as a foster-mother, and that some day he will know more of the true state of affairs when he visits the smithy at the cross-roads.

Another circumstance to which the badger hereabouts is indebted for security is that terriers are not the favourite dogs of the countryside. When shooting, the sportsman prefers spaniels, particularly certain "strains" of black and brown cockers—untiring little workers with a keen, true power of scent—which for many years have been common in the neighbourhood; and the farmer's sheep-dog is unfitted for any sport except rabbiting. Here and there, among the poaching fraternity, may be found a mongrel fondly imagined by its owner to be a terrier—a good rabbit "marker," and wonderfully quick in killing rats, but no more suited than the sportman's spaniel for "lying up" with a badger.

Undoubtedly, however, the security of some of our most interesting wild animals, and especially of the badger, is to be accounted for by their extreme shyness. They venture abroad only when the shadows of night lie over the woods. For countless years, dogs and men have been their greatest foes, and their fear of them is found to be almost as strong in remote districts as where, near towns, their existence is continually threatened. Wild [Pg 298] life in our quiet valley will be deemed of unusual interest when I say that less than six hours before writing these lines I visited a badger's "set"—a deep underground hollow with several main passages and upper galleries, where, as I have good reason to believe, a fox also dwells—an otter's "holt" beneath gnarled alder-roots fringing the river-bank, and another fox's "earth," all on the outskirts of a wooded belt not more than a mile from my home, and all showing signs of having long been inhabited.

Unless systematically persecuted, the fox, the otter, and the badger cling to their respective haunts with such tenacity that, season after season, they prowl along the same familiar paths through the woods or by the river, and rear their young in the same retreats. This is the case especially with the badger; from the traditions of the countryside, as well as from the careful observation of sporting landowners, it may be learned that for generations certain inaccessible "sets" have seldom, if ever, been uninhabited. Always at nightfall the "little man in grey" has climbed the [Pg 299] slanting passage from his cave-like chamber, ten or—if among the boulders of some ancient cairn—even from twenty to thirty feet below the level of the soil, and sniffed the cool evening air, and listened intently for the slightest sound of danger, before departing on his well worn trail to hunt and forage in the silent upland pastures. And with the first glimpse of light, when the hare stole past towards her "form," and the fox, a shadowy figure drifting through the haze of early dawn, returned to the dense darkness of the lonely wood, he has sought his daytime snuggery of leaves and grass industriously gathered from the littered glades.

In a deep burrow at the foot of a hill, about a quarter of a mile from a farmstead built on a declivity at a bend of the broad river, Brock, the badger, was born, one morning about the middle of spring. Three other sucklings, like himself blind and wholly dependent on their parents' care, shared his couch of hay and leaves. Day by day, the mother badger, devoted to their welfare, fed and tended her unusually numerous offspring, lying beside them on the comfortable [Pg 300] litter, while the sire, occupying a snug corner of the ample bed, dozed the lazy hours away; and evening after evening, when twilight deepened into darkness as night descended on the woods, she arose, shook a few seed-husks from her coat, and with her mate adjourned to an upper gallery leading to the main opening of the "set," whence, assured that no danger lurked in the neighbourhood of their home, both stole out to forage in the clearings and among the thickets on the brow of the hill.

Just as with Lutra, the little otter-cub in the "holt" above the river's brim, the first weeks of babyhood passed uneventfully, so with Brock, the badger, nothing of interest occurred till his eyes gradually opened, and he could enjoy with careless freedom the real beginning of his woodland life. Even thus early, what may be called the nocturnal instinct was strong within him. He was alert and playful chiefly at night, when, deep in the underground hollow, nothing could be heard of the outer world but the indistinct, monotonous wail of the wind in the upper passages of the "set." Droll, indeed, were the revels of the young badgers [Pg 301] when the parents were hunting far away. The little creatures, awakened from a heavy sleep that had followed the last fond attentions of their mother, were loath to frolic at once with each other in the lonely, silent chamber. In their parents' absence they felt unsafe; that mysterious whisper of admonition, unheard but felt, which is the voice of the all-pervading spirit of the woods forever warning the kindred of the wild, bade them be quiet till the dawn should bring the mother badger to the lair once more. So, huddled close, they were for a time satisfied with a strangely deliberate game of "King of the Castle," the castle being an imaginary place in the middle of their bed. Towards that spot each player pushed quietly, but vigorously, one or other gaining a slight advantage now and again by grunting an unexpected threat into the ear of a near companion, or by bestowing an unexpected nip on the flank of the cub that held for the moment the coveted position of king. Withal this was a sober pastime, unless Brock, the strongest and most determined member of the family, chanced to provoke his playmates [Pg 302] beyond endurance, and caused a general, reckless scramble, in which tiny white teeth were bared and tempers were uncontrolled.

As the night wore on, it almost invariably happened, however, that the "Castle" game gave place to a livelier diversion akin to "Puss in the Corner," when, on feeble, unsteady legs, the "earth-pigs" romped in pursuit of each other, or squatted, grunting with excitement, in different spots near the wall of their nursery. But, tired at last, they ceased their gambols an hour or so before dawn, lay together in a warm, panting heap, and slept, till, on the return of their mother to the "set," they were gathered to the soft comfort of her folded limbs, and fed and fondled to their hearts' content.

Though Brock grew as rapidly as any young badger might be expected to grow, a comparatively long time passed by before he and the other small members of the family ventured out of doors. Repeatedly they were warned, in a language which soon they perfectly understood, that, except under the care of their parents, a visit to the outer world would end disastrously; so, while the [Pg 303] old ones were abroad, the little creatures dared not move beyond the opening to the dark passage between the chamber and the gallery above. Sometimes, following their dam when she climbed the steep passage to her favourite lookout corner within a mouth of the burrow, they caught a glimpse of the sky, and of the trees and the bracken around their home; but a journey along the gallery was never made before the twilight deepened.

The purpose of such close confinement was, that the young badgers should be taught, thoroughly and without risk, the first principles of wood-craft, and thus be enabled to hold their own in that struggle for existence, the stress of which is known even to the strong. Obedience, ever of vital importance in the training of the forest folk, was impartially exacted by the mother from her offspring. It was also taught by a system of immediate reward. The old badger invariably uttered a low but not unmusical greeting when she returned to her family at dawn. Almost before their eyes were open, the sucklings learned to connect this sound with food and comfort, and at once turned to the spot from [Pg 304] which it proceeded. Later, when the same note was used as a call, they recognised that its meaning was varied; in turn it became, with subtle differences of inflection, an entreaty, a command, and a warning that it would be folly to ignore; but, whatever it might indicate, they instinctively remembered its first happy associations, and hurried to their mother's side. Hardly different from the call, when it conveyed the idea of warning, was a note of definite dissent, directing the youngsters to cease from squabbling, and to become less noisy in their rough-and-tumble play. After they had learned each minute difference in the call notes, their progress in education was largely determined by that love of mimicry which always prompts the young to imitate the old; and in time they acquired the tastes, the passions, and the experiences of their watchful teachers.

While prevented from wandering abroad, they nevertheless were not entirely ignorant of what was happening in the woods. They were not quickly weaned; it was necessary, before the dam denied them Nature's first nourishment, that they should have ready access to the brook that trickled down the [Pg 305] hillside hollow not far from the "set." But meanwhile, young rabbits, dug from the breeding "stops" of the does, were frequently brought to them, and the badgers were encouraged to gratify a love for solid food which nightly became stronger.

In this part of the education of their young, the parent badgers adopted methods similar to those of the fox and other carnivorous animals. When first the mother badger brought a rabbit home, she placed it close beside her cubs, so that they could not fail to be attracted by its scent. For a moment, aware of something new and strange, they showed signs of timidity, and crouched together in the middle of the nest; but the presence of their mother reassured them, and they sniffed at the warm body with increasing delight. The dam seemed to know each trifling thought passing through their minds;

and, observing their eager interest, she dragged the rabbit into a corner of the bed, making great show of savagery, as if guarding it from their attacks. Time after time, she alternately surrendered and withdrew her victim, till the tempers of the little animals, irritated beyond control by her tantalising methods, blazed out in a free fight among [Pg 306] themselves for possession of the prize. The mother now retired to a corner of the "set," and listened attentively to all that happened, till they had finished their quarrel, and Brock, the middle figure in a group of tired youngsters, lay fast asleep with his head on the rabbit's neck. Then she turned, climbed quietly to the upper galleries, and, stealing out among the shadows of the wood, came again to the breeding "stop," where she unearthed and devoured a young rabbit that had been suffocated in the loose soil thrown up during her former visit. After quenching her thirst at the brook in the hollow, she journeyed to the upland fields, crossed the scent of her mate in the gorse, and then "cast" back across the hillside, making a leisurely examination of each woodland sign, to satisfy herself that no danger lurked in the neighbourhood of her home.

For the badger, as for the tiny field-vole in the rough pastures of the Cerdyn valley, the various scents and sounds were full of meaning, and constituted a record of the night such as only the woodland folk have learned fully to understand. The [Pg 307] smell of the fox lay strong on a path between the oaks; with it was mingled the scent of a bird; and a white feather, caught by a puff of wind, fluttered in the grass: young Reynard, boldest of an early family in the "earth," had stolen a fowl from a neighbouring farmyard near the river, and had carried it—not slung over his shoulders, as fanciful writers declare, but with its tail almost touching the soil—into the thicket beyond the wood. Rabbits had wandered in the undergrowth; and, near a large warren, the stale, peculiar odour of a stoat that had evidently prowled at dusk lingered on the dewy soil. The signs of blackbirds and pigeons among the loose leaf-mould were also faint; as soon as night had fallen, the birds had flown to roost in the branches overhead. The short, coughing bark of an old fox came from the edge of the wood; and then for some time all was quiet, till the musical cry of an otter sounded low and clear from the river beneath the steep.

These familiar voices of the wilderness caused the badger no anxiety; they told her of freedom from danger; they were to [Pg 308] her assuring signals from the watchers of the night. But the howl of a dog in a distant farmstead, and the bleat of a restless sheep in the pasture on the far side of the hill, told her a different story; they reminded her, as the smell of the fowl had done, that man, arch-enemy of the woodland people, might in any capricious moment threaten her existence, seeking to destroy her even while by day she slumbered in her chamber under the roots of the forest trees.

She crossed the gap, where the river-path joined the down-stream boundary of the wood, then, with awkward, shambling stride, climbed the steep pasture, and for a few moments paused to watch and listen in the deep shadows of the hedge on the brow of the slope. A rabbit, that had lain out all night in her "seat" beneath the briars, rushed quickly from the undergrowth, and fled for safety to a burrow in the middle of the field. A small, dim form appeared for a moment by a wattled opening between the pasture and the cornfield above, then, with a rustle of dry leaves, vanished on the further side—a polecat was returning to her [Pg 309] home in a pile of stones that occupied a hollow on the edge of the wood.

Day was slowly breaking. A cool wind, blowing straight from the direction of a homestead indistinctly outlined against the dawn, stirred the leaves in the ditch, and brought to the badger's nostrils the pungent scent of burning wood—the milkmaid was already at work preparing a frugal breakfast in the kitchen of a lonely farm. Fearing that with the day the birds would mock her as she passed, and thus reveal her whereabouts to some inquisitive foe, the badger sought the loneliest pathway through the wood, and returned, silently but hastily, to her home.

II.

HOME DISCIPLINE.

During the mother badger's absence from home, an unlooked-for event—almost the exact repetition of an incident in the training of Vulp, the young fox—had happened in the education of her cubs. Her mate, hunting in an upland fallow, had been surprised by a poacher, and, long before daybreak, had discreetly returned to the "set." The success he had met with had enabled him to feed to repletion, so he was not tempted by the dead rabbit carried home by the mother and left in the chamber. Fearing to leave his hiding place, he wisely determined to devote the time at his disposal, before settling to sleep, to his children's instruction. With a grunt like that of the mother when she greeted her offspring, he at once aroused the slumbering youngsters, and then, heedless of their attentions, as, mistaking him for the dam, they pressed at his side, he laid hold of the rabbit and dragged it into a far corner. Full of curiosity, the cubs followed, but with well assumed anger he drove them away. As if in keen anticipation of a feast, he tore the dead animal into small pieces which he placed together on the floor of the chamber. This task complete, he retired to his accustomed resting place, and listened while the cubs, overcoming their timidity, ventured nearer and nearer to the dismembered rabbit, till, suddenly smelling the fresh blood, they gave way to inborn passion, and buried their teeth in the lifeless flesh. An inevitable quarrel ensued; Brock and his companions could not agree on the choice of tit-bits, and a medley of discordant grunts and squeals seemed to fill the chamber, though now and again it partly subsided, as two or three of the cubs, having fixed on the same portion of the rabbit, tugged and strained for its possession—so intent on the struggle that they dared not waste their breath in useless wrangling.

The old badger, satisfied that his progeny gave excellent promise of pluck and strength, was almost dropping off contentedly to sleep, when one of the excited combatants, retreating from the fray, backed unceremoniously, and awoke him with an accidental blow on the ribs. This was more than the crusty sire could endure, and he administered such prompt and indiscriminate chastisement to the youngsters, that, in a subdued frame of mind, they forgot their differences, forgot also the toothsome remnants of their feast, and nestled together in bed, desiring much that their patient dam would come to console them for the ill-usage just received.

On returning to the "set," the mother badger stayed for a few minutes at the edge of the mound before the main entrance, and, rearing herself on her hind-legs, rubbed her cheek against a tree-trunk, and sniffed the air for the scent of a lurking enemy. Then, satisfied that all was safe, she entered the deep chamber, and was greeted by the little creatures that for an hour had expectantly awaited her arrival. Unusually boisterous in their welcome, they instantly disregarded the presence of their sire; and [Pg 313] such, already, was the magic effect of the meal of raw flesh on their tempers, that, with an eagerness hitherto unknown, they followed every movement of their dam, till, submitting to their importunities, she lay beside them, and fed and fondled them to sleep.

Almost nightly, she brought something new with which to tempt their appetites—young bank-voles dug from their burrows on the margin of the wood, weakling pigeons dropped from late nests among the leafy boughs, snakes, and lizards, and, chiefly, suckling rabbits unearthed from the shallow holes which the does had "stopped" with soil thrown back into the entrance when they left to feed amid the clover.

Though young rabbits, in breeding "stops" barely a foot below the level of the ground, were never safe from the badger's attack, a flourishing colony dwelt within the precincts of the "set." Early in spring, when the badgers were preparing for their expected family, a doe rabbit, attracted by the great commotion caused by their efforts to remove the big heap of soil thrown up at the entrance to their dwelling, hopped quietly out of the [Pg 314] fern, and sat for a long time watching from between the bushes the occasional showers of loam which indicated the progress of the work. Judged by the standard of a rabbit, Bunny was a fairly clever little creature, and the plans she formed as she hid in the undergrowth seemed to show that she possessed unusual forethought. She waited and watched for several nights, till the badgers had ceased to labour, and the mound before the "set" remained apparently untouched. Then, one evening, after she had seen the badgers go off together into the heart of the wood, she entered, and moved along the gallery, pausing here and there to touch the walls with her sensitive muzzle. Coming to a place where a stone was slightly loosened, she began to dig a shaft almost at right angles to the roomy gallery, and for a time continued her work undisturbed; but an hour or so before dawn she retired to sleep in a thicket, some distance beyond the plain, wide trail marking the badger's movements to and from the nearest fields.

The badgers, on returning home, were sorely puzzled at the change that had taken [Pg 315] place during their absence. To all appearance, a trick had been played on them, for, whereas their house had been left neat and tidy at dusk, there was now a pile of earth obstructing the main passage. However, they accepted the situation philosophically, and completed the rabbit's work by clearing the gallery and adding to the heap beyond the entrance.

Night after night, the wily rabbit watched for the badgers' departure, carried on her work, and gave them a fresh task for the early morning, till a short but winding burrow, some depth below the level of the ground, formed an antechamber where the little family to which she presently gave birth was reared in safety.

Though the badgers, aware that the shallow "stops" in the woods were more easily unearthed than this deeper burrow near the mouth of the "set," did not seek to disturb their neighbours, the mother rabbit, directly her family grew old enough to leave the nest, became increasingly vigilant, and, when about to lead them to or from their dwelling, was ever careful to be satisfied that all was quiet in the chambers and the galleries below. Generally she ventured abroad before the [Pg 316] badgers awoke from the day's sleep, came back during their absence, and once more stole out to feed when they had returned and were resting in their snuggery. The danger that lurked in her surroundings supplied a special excitement to life, and she never heard without fear the ominous sounds that vibrated clearly through every crack and cranny when the badgers occasionally arose from their couch, stretched their cramped limbs, shook their rough grey coats, and grunted with satisfaction at the feeling of health and strength which nearly all wild animals delight occasionally to express.

The forest trees had donned their verdure; the tall bracken had lifted its fronds so far above the grass that the mother rabbit no longer found them a convenient screen through which to peer at the strange antics of the old badgers as they came from their lair and sat in the twilight on the mound by the entrance of their home; and the rill in the dingle, which, during winter and early spring, leaped, a clear, rushing torrent, on its way to the river below the steep, had dwindled to a few drops of water, collected in tiny pools among the stones, or trickling reluctantly [Pg 317] down the dank, green water-weed. The young badger family had grown so strong and high-spirited that their dam, weakened by motherhood, and at a loss to restrain their increasing desire for outdoor air and exercise, determined to wean them, and to teach them many lessons, concerning the ways of the woodland people, which she had learned long ago from her parents, or, more recently, from her own experiences as a creature of the dark, mysterious night.

Brock, in particular, was the source of considerable anxiety to her. He was the leader in every scene of noisy festivity; she was repeatedly forced to punish him for following her at dusk to the mound outside the upper gallery, and for disobedience when she condescended to take part in a midnight romp in the underground nursery. He tormented the other members of the family by awakening them from sleep when he desired to play, also by appropriating, till his appetite was fully appeased, all the food his dam brought home from her hunting expeditions, and, again, by picking quarrels over such a trifling matter as the choice of a place when [Pg 318] he and his little companions wished to rest.

Nature's children are wilful and selfish; and in their struggle for existence they live, if independent of their parents, only so long as they can take care of themselves. Among adult animals, however, selfishness seems to become inoperative in the care they take of their offspring. But though the mother badger was unselfish towards her little ones, she spared no effort to instruct them in the ways of selfishness.

The night of Brock's first visit to the woods was warm and unclouded. For an hour after sunset, he played about the gallery by the door, while his mother, a vigilant sentinel, remained motionless and unseen in the darkness behind. Now and again, he heard the rabbits moving in the burrow, but they, aware of his presence, stayed discreetly out of view. Under his mother's guidance, or even if his playmates had been bold enough to accompany him, he would at once have been ready to explore the furthest corner of the rabbit-hole. But the old badger was too big, and the youngsters were too timid, to go with him [Pg 319] into the mysterious antechamber; so, after repeated attempts to explore the passage as far as the bend, and finding to his discomfort that there the space became narrower, he gave up the idea of prying on the doings of his neighbours, and contented himself with droll, clumsy antics, such as those by which wild children often seek to convince indulgent parents that they are eager and fearless.

As the darkness deepened, the dog-badger, after hunting near the outskirts of the wood, returned to the "set." His manner indicated that he was the bearer of an important message. He touched his mate on the shoulder; then, as she responded to his greeting, he thrust his head forward so that she could scent a drop of blood clinging to his lip; and, while she sniffed enquiringly along the fringe of his muzzle, he seemed to be assuring her that his message was of the utmost consequence. As soon as she understood his meaning, he vanished into the gallery, and for a few moments was evidently busy. Faint squeals and grunts, which gradually became louder and louder, proceeded from the central [Pg 320] chamber, and, again, from the inner passages; and presently the big badger appeared in sight, driving his family before him, and threatening them with direst punishment if they attempted to double past him and thus regain their dark retreat.

Wholly unable to appreciate the real position of affairs, Brock, perplexed and frightened, found himself hiding among the ferns and brambles outside the "set," while the sire, standing in full view on the mound, and grunting loudly, forbade the return of his evicted family. Unexpectedly, too, the mother badger, when the little ones looked to her for sympathy in their extraordinary treatment, took the part of the crusty old sire, and snapped and snarled directly they attempted to move back towards the mound. Utterly bewildered and much in fear, since their dam, hitherto the object of implicit trust, had suddenly deserted their cause, the young badgers crouched together under the bushes, and watched distrustfully each movement of their parents. The sire stuck to his post on the mound, and, with hoarse grunts, varied occasionally by thin, piping squeals that did not seem in the [Pg 321] least to accord with his wrathful demeanour, continued to keep them at a distance.

Soon the dam moved slowly away, climbed the track towards the top of the wood, and then called to the cubs as they sat peering after her into the darkness. Released from discipline, and eagerly responsive to her cry, they lurched after her, and followed closely as she led them further and still further from home. Presently, the dog-badger overtook his family. His manner, as well as the dam's, had changed; and though great caution was exercised as they journeyed along paths well trodden, and free from twigs that might snap, or leaves that might rustle, and though silence was the order of the march, the little family—proud parents and shy, inquisitive children—seemed as happy as the summer night was calm. The distant sound of a prowling creature, heard at times from the margin of the wood, caused not the slightest alarm to the cubs: the intense nervousness always apparent in young foxes was not evinced by the little badgers.

In comparison with the fox-cubs, they were not easily frightened; they already [Pg 322] gave promise of the presence of mind which, later, was often displayed when they were threatened by powerful foes. Brock, nevertheless, betrayed astonishment when a dusky form bolted through the whinberry bushes close by; and several moments passed before he was able, by his undeveloped methods of reasoning, to connect the scent of the flying creature with that of the rabbits often carried home by his mother, and, therefore, with something good for food.

At the top of the wood, the old badgers turned aside and led the way through a thicket, where, in obedience to their mother, the youngsters came to a halt, while their sire, proceeding a few yards in advance, sniffed the ground, like a beagle picking up the line of the hunt. Having found the object of his search, he called his family to him, that they might learn the meaning of the various signs around. But the doings of the woodland folk could not yet be learnt by the little badgers, as by the experienced parents, from trifling details, such as the altered position of a leaf or twig, the ringing alarm-cry of a bird, the fresh earth-smell near an upturned [Pg 323] stone, or the taint of a moving creature in the grass. Beside them lay a small brown and white stoat, its head almost severed from its body by a quick, powerful bite, and, just beyond, the motionless form of a half-grown rabbit, unmarked, save by a small, clean-cut wound between the ears. The scent of both creatures was noticeable everywhere around, and with it, quite as strong and fresh, the scent of the big male badger. Walking up the path, soon after nightfall, the badger had arrived on the scene of a woodland tragedy, and had found the stoat so engrossed with its victim that to kill the bloodthirsty little tyrant was the easy work of an instant. Afterwards, mindful of the education of his progeny, he had hurried home to arrange with his mate a timely object lesson in wood-craft.

The stoat was left untasted, but the rabbit was speedily devoured; and then the badger family resorted to the riverside below the "set," where the cubs were taught to lap the cool, clear water. Thence, before returning home, they were taken to a clearing in the middle of the wood, and, while the sire [Pg 324] went off alone to scout and hunt, the mother badger showed them how to find grubs and beetles under the rotting bark of the tree-butts, in the crevices among the stones, and in the soft, damp litter of the decaying leaves.

III.

FEAR OF THE TRAP.

Night after night, the cubs, sometimes under the protection of both their parents, and sometimes under the protection of only the dam, roamed through the by-ways of the countryside. From each expedition they gleaned something of new and unexpected interest, till they grew wise in the ways of Nature's folk that haunt the gloom—the strong, for ever seeking opportunities of attack; the weak, for ever dreading even a chance shadow on the moonlit trail.

A strange performance, which, for quite a month, seemed devoid of meaning to the cubs, but which, nevertheless, Brock soon learned to imitate, took place whenever the tainted flesh of a dead creature was found in the way. The old badgers at once became alert, moved with the utmost caution, smelt but did not touch the offensive morsel, and, instead of seizing it, rolled over it again and yet again, as if the scent proved irresistibly attractive. One of the cubs, that had always shown an inclination to act differently from the way in which her companions acted, and often became lazy and stupid when lesson-time arrived, was destined to pay dearly for neglecting to imitate her parents. Lagging behind the rest of the family, as in single file they moved homeward after a long night's hunting in the fallow, she chanced to scent some carrion in the ditch, turned aside to taste it, and immediately was held fast in the teeth of an iron trap. Hearing her cries of pain and terror, the mother hastened to the spot, and, for a moment, was so bewildered with disappointment and anger that she chastised the cub unmercifully, though the little creature was enduring extreme agony. But directly the old badger recovered from her fit of temper, she sought to make amends by petting and soothing the frightened cub, and trying to remove the trap. Finally, after half an hour's continuous effort, she accidentally found that the trap was connected by a chain with a stake thrust into the ground. Quickly, with all the strength of her muscular fore-paws, she dug up the soil at the end of the chain, and then, with powerful teeth, wrenched the stake from its position. Dragging the cruel trap, the young badger slowly followed her dam homeward, but when she had gone about a hundred yards pain overcame her, and she rolled down a slight incline near the hedge. For a few minutes, she lay helpless; then, grunting hoarsely, she climbed the ditch, and continued her way in the direction of a gap leading into the wood. There, as she gained the top of the hedge, the trap was firmly caught in the stout fork of a thorn-bush. Further progress was impossible; all her frantic struggles failed to give her freedom. The dam stayed near, vainly endeavouring to release her, till at dawn a rustle was heard in the hedge, and a labourer on his way to the farm came in sight above a hurdle in the gap. Reluctantly, the old badger stole away into the wood, leaving the cub to her fate. It came—a single blow on the nostrils from a stout cudgel—and all was over. [Pg 328]

The lesson thus taught left a salutary impression on the minds of the other cubs. From it they learned that the presence of stale flesh was somehow associated with the peculiar scent of oiled and rusty iron, or with the taint of a human hand, and was fraught with the utmost danger. They somehow felt that their dam acted wisely in rolling over any decaying refuse she happened to find on her way; and later, when Brock, seizing an opportunity to imitate his mother, sprang another trap, which, closing suddenly beneath his back, did no more harm than to rob him of a bunch of fur, they recognised how a menace to their safety might be easily and completely removed by the simple expedient taught them by their careful parent.

Though she invariably took the utmost precaution against danger from baited traps, the old she-badger was nevertheless surprised, almost as much as were the cubs, at the incidents just described. At various times she had sprung more than a dozen traps, but in each case her attention had been directed to the trap only by the scent of iron, or of the human hand. However faint that scent might be, and however mingled [Pg 329] with the smell of newly turned earth or of sap from bruised stalks of woodland plants, she immediately detected it, rolled on the spot, and then noted the signs around—the disturbed leaf-mould, and the foot-scent of man leading back among the bilberry bushes, or down the winding paths between the oaks, where, occasionally, she also found faint traces of the hand-scent on bits of lichen, or on rotten twigs, fallen from the grasp of her enemy as he clutched the tree-trunks in his steep descent towards the riverside. But never before had she seen a baited trap. Her dam had never seen one; her grand-dam had been equally ignorant; and yet both, like herself, had always rolled on any tainted flesh they chanced to come across on their many journeys.

For generations, in this far county of the west, the creatures of the woods, except the fox, had never been systematically hunted. The vicissitudes of history had directly affected the welfare of wild animals. The old professional hunting and fighting classes had become unambitious tenant farmers; and, partly through the operations of an old Welsh law regarding the equal [Pg 330] division of property, the land beyond the feudal tracts of the Norman Marches were, in many instances, broken up into small freeholds owned by descendants of the princely families of bygone ages. But hard, incessant work was the lot of tenant and freeholder alike. When the aims and the experiences of the old fighting and sporting days had passed away, and nothing was left but ceaseless toil, these essentially combative people, to whom violent and continuous excitement was the very breath of life, became, for a while at least, knavish and immoral, sunk almost to one dead social level, and totally uninteresting because, in their new life of peaceful tillage—a life far more suited to their English law-givers than to themselves—they were apparently incapable of maintaining that complete, vigilant interest in their ordinary surroundings which makes for enlightenment and success.

Having lost the love of "venerie" possessed by their forefathers, the farmers cared little about any wild creatures but hares and rabbits; a badger's ham was to them an unknown article of food. The fear [Pg 331] of a baited trap had, therefore, probably descended from one badger to another since days when the green-gowned forester came to the farm, from the lodge down-river, and sought assistance in the capture of an animal for the sport of an otherwise dull Sunday afternoon in the courtyard of the nearest castle; or even since ages far remote, when a badger's flesh was esteemed a luxury by the earliest Celts.

Unbaited traps, in the "runs" of the rabbits, had at intervals been common for centuries; but now the carefully prepared baits and the unusually strong traps seemed to indicate nothing less than an organised attack on other and more powerful night hunters. The badger's fears, however, were hardly warranted. Five traps had been placed in the wood by a curious visitor staying at the village inn. In one of these, Brock's sister had been caught; but the owner of the trap knew nothing beyond the fact that it had mysteriously disappeared from the spot where he had seen it fixed. Another was sprung by Brock; two at the far end of the wood were so completely fouled by a fox that every prowling creature carefully avoided [Pg 332] the spot; while in the fifth was found a single blood-stained claw, left to prove the visit of a renegade cat.

It may well be imagined that a large and interesting animal like the badger, keeping for many years to an underground abode so spacious that the mound at its principal entrance is often a quite conspicuous landmark for some distance in the woods, would be subject to frequent and varied attacks from man, and thus be speedily exterminated. It may also be imagined that the habits of following the same well worn paths night after night, of never ranging further than a few miles from the "set," and of living so sociably that the community sometimes numbers from half-a-dozen to a dozen members, apart from such lodgers as foxes, rabbits, and wood-mice, would all combine to render the creature an easy prey.

But if the badger's ways are carefully studied, the very circumstances which at first seem unfavourable to him are found to account for much of his immunity from harm. The depth of his breeding chamber and the length of the connecting passages are, as a rule, indicated by the size [Pg 333] of the mound before his door. The fact that he regularly pursues the same paths in his nightly excursions enables him to become familiar, like the fox, with each sight and scent and sound of the woods, so that anything strange is at once noticed, and danger avoided. His sociability is a distinct gain, because he receives therefrom co-operation in his sapping and mining while he aims to secure the impregnability of his fortress; and his tolerance of cunning and timid neighbours gains for him this advantage: sometimes in the dusk, before venturing abroad, he receives a warning that danger lurks in the thickets around his home—perhaps from a double line of scent indicating that the fox has started on a journey and then hurriedly turned back, or from numerous cross-scents at the mouth of the burrow, where the rabbits and the wood-mice have passed to and fro, deterred by fear in their frequent attempts to reach feeding places beyond the nearest briar-clumps. His methods, however, when either his neighbours or the members of his own family become too numerous, are prompt and drastic. [Pg 334]

Shy, inoffensive, and, for a young creature unacquainted with the responsibilities of a family, deliberate to the point of drollery in all his movements, Brock grew up beneath his parents' care; and, with an intelligence keener than that possessed by the other members of the little woodland family, learned many lessons which they failed to understand. When his mother called, he was always the first to hasten to her side. Each incident of the night, if of any significance, was explained to her offspring by the mother. Often Brock was the only listener when she began her story, and the late arrivals heard but disconnected parts.

Beautiful beyond comparison were those brief summer nights, silent, starlit, fragrant, when the badgers led their young by many a devious path through close-arched bowers amid the tangled bracken, or under drooping sprays of thorn and honeysuckle in the hidden ditches, or through close tunnels, as gloomy as the passages of their underground abode, in the dense thickets of the furze. Sometimes they wandered in the corn and root-crop, or in the hayfield [Pg 335] where the sorrel, a cooling medicinal herb for many of the woodland folk, grew long and succulent; and sometimes they descended the steep cattle-path on the far side of the farm, where the big dor-beetles, as plentiful there as in the grass-clumps of the open pasture, were easily struck down while they circled, droning loudly, about the heaps of refuse near the hedge.

Once, late in July, when the badgers were busily catching beetles by the side of the cattle-path, Brock, thrusting his snout into the grass to secure a crawling insect, chanced to hear a faint, continuous sound, as of a number of tiny creatures moving to and fro in a hollow beneath the moss-covered mound at his feet. He listened intently, his head cocked knowingly towards the spot whence the sound proceeded; then, scratching up a few roots of the moss, he sniffed enquiringly, drawing in a long, deep breath, at the mouth of a thimble-shaped hole his sharp claws had exposed.

Unexpectedly, and without the help of the dam, he had discovered a wild-bees' nest. His inborn love of honey was every whit as strong as a bear's, and he recognised the [Pg 336] scent as similar to that of insects known by him to be far more tasty than beetles; so, without a moment's hesitation, he began to dig away the soil. The nest was soon unearthed, and the little badger, completely protected by his thick and wiry coat from the half-hearted assaults of the bewildered bees, greedily devoured the entire comb, together with every well-fed grub and every drop of honey the fragile cells contained. His eagerness was such that these spoils seemed hardly more than a tempting morsel sufficient to awaken a desire for the luscious sweets of the wayside storehouses. He carefully hunted the hedgerow, as far as a gate leading to a rick-yard, and at last, close to a stile, found another nest, which, also, he quickly destroyed.

Henceforth, till the end of August, there were few nights during which he did not find a meal of honey and grubs. The summer was fine and warm, a lavish profusion of flowers adorned the fields and the woods, and humble-bees and wasps were everywhere numerous. As if to taunt the badgers with inability to climb, a swarm of tree-wasps lived in a big nest of wood-pulp [Pg 337] suspended from a branch ten feet or so above the "set," and, every afternoon, the badgers, as they waited near the mouth of their dwelling for the darkness to deepen, heard the shrill, long continued humming of the sentinel wasps around the big ball in the tree—surely one of the most appetising sounds that could ever reach a badger's ears. But the wasps that had built among the ferns near the river-path, and in the hollows of the hedges, were remorselessly hunted and despoiled. Their stings failed to penetrate the thick coat and hide of their persistent foes, while a chance stab on the lips or between the nostrils seemed only to arouse the badgers from leisurely methods of pillage to quick and ruthless slaughter of the adult insects as well as of the immature grubs. But Brock never committed the indiscretion of swallowing a full-grown wasp. With his fore-paws he dexterously struck and crippled the angry sentinels that buzzed about his ears, and, with teeth bared in order to prevent a sting on his tender muzzle, disabled the newly emerged and sluggish insects that wandered over the comb. [Pg 338]

As autumn drew on, the cubs grew strong and fat on the plentiful supplies of food, which, with their parents' help, they readily found in field and wood. Brock gave promise of abnormal strength, and was already considerably heavier than his sister. They fared far better than the third cub, a little male, that, notwithstanding a temper almost as fiery as Brock's, was worsted in every dispute and frequently robbed of his food, and still, never owning himself beaten, persisted in drawing attention to his success whenever he happened on something fresh and toothsome. At such times, instead of hastily and silently regaling himself, he made a great a-do, grunting with rage and defiance, like a dog that guards a marrow-bone but will not settle down to gnaw its juicy ends.

Brock's brother was so often deprived of his legitimate spoils, that, while his surliness was increased, his bodily growth was checked. He was small and thin for his age; and so, when a kind of fever peculiar to young badgers broke out in the woodland home, he succumbed. His grave was a shallow depression near the path below the "set," whither his parents dragged his lifeless body, [Pg 339] and where the whispering leaves of autumn presently descended to array him in a red and golden robe of death.

The other young badgers quickly recovered from their fever; and by the end of October all the animals were, as sportsmen say, "in grease," and well prepared for winter's cold and privation. The old badgers became more and more indisposed to roam abroad; and, whereas in summer they sometimes wandered four or five miles from the "set," they now seldom went further than the gorse-thicket on the fringe of the wood.

IV.

[Pg 340]
Top

THE WINTER "OVEN."

The badger-cubs, while not so well provided against the cold as were their parents, grew lazy as winter advanced, and spent most of their time indoors on a large heap of fresh bedding, that had been collected under the oaks and carried to a special winter "oven" below the chamber generally occupied in summer. Here, the sudden changes of temperature affecting the outer world were hardly noticeable; and so enervating were the warmth and indolence, that the badgers, in spite of thick furs and tough hides, rarely left their retreat when the shrill voice of the north-east wind, overhead in the mouth of the burrow, told them of frost and snow.

About mid-winter, the first of two changes took place in the colour of the young badgers' [Pg 341] coats; from silver-grey it turned to dull brownish yellow, and the contrasts in the pied markings of the cheeks became increasingly pronounced. This change happened a little later with Brock than with his sister. Eventually, late in the following winter, the young female, arriving at maturity, donned a gown of darker grey, and her face was striped with black and white; shortly afterwards, Brock, too, assumed the livery of a full-grown badger.

Meanwhile, till events occurred of which the second change was only a portent, all remained fairly peaceful in the big burrow under the whins and brambles. Occasionally, in the brief winter days, Brock was awakened from his comfortable sleep by the music of the hounds, as they passed by on the scent of Vulp, the fleetest and most cunning fox on the countryside, or by the stamp of impatient hoofs, as the huntsman's mare, tethered to a tree not far from the "set," eagerly awaited her rider's return from a "forward cast" into the dense thicket beyond the glade.

One afternoon in late winter, a young vixen, that, without knowing it, had completely [Pg 342] baffled her pursuers, crept, footsore and travel-stained, into the mouth of the "set," and lay there, panting loudly, till night descended, and she had sufficiently recovered from her distress to continue her homeward journey. Now and again, the sharp report of a shotgun echoed down the wood; and once, late at night, when Brock climbed up from the "oven" to sit awhile on the mound before his door, the scent of blood was strong in the passage leading to the rabbit's quarters. Unfortunate bunny! Next night, stiff and sore from her wounds, she crawled out into the wood, and Vulp and his vixen put an end to her misery long before the badgers ventured from their lair.

Winter, with its long hours of sleep, passed quietly away. Amid the sprouting grasses by the river-bank, the snowdrops opened to the breath of spring; soon afterwards, the early violets and primroses decked the hedgerows on the margin of the wood, and the wild hyacinths thrust their spike-shaped leaves above the mould. The hedgehogs, curled in their beds amid the wind-blown oak-leaves, were awakened by the gentle heat, and wandered through [Pg 343] the ditches in search of slugs and snails. One evening, as the moon shone over the hill, the woodcock, that for months had dwelt by day in the oak-scrub near the "set," and had fed at night in the swampy thickets by the rill, heard the voice of a curlew descending from the heights of the sky, and rose, on quick, glad pinions, far beyond the soaring of the lark, to join a great bird-army travelling north. Regularly, as the time for sleep drew nigh, the old inhabitants among the woodland birds—the thrushes, the robins, the finches, and the wrens—squabbled loudly as they settled to rest: their favourite roosting places were being invaded by aliens of their species, that, desirous of breaking for the night their northward journey, dropped, twittering, into every bush and brake on the margin of the copse. And into Nature's breast swept, like an irresistible flood, a yearning for maternity.

The vixen, that once had rested inside the burrow to recover from her "run" before the hounds, remembered the sanctuary, returned to it, and there in time gave birth to her young; and, though almost [Pg 344] in touch with such enemies as the badger and the fox, a few of the rabbits that had been reared during the previous season in the antechamber of the "set" enlarged their dwelling place, and were soon engaged in tending a numerous offspring. The timid wood-mice, following suit, scooped out a dozen tiny galleries within an old back entrance of the burrow, and multiplied exceedingly. But, while all other creatures seemed bent on family affairs, Brock's parents, following a not infrequent habit of their kindred, deferred such duties to another season.

As spring advanced, food became far more abundant than in winter, and the badgers' appetites correspondingly increased. Directly the evening shadows began to deepen, parents and cubs alike became impatient of the long day's inactivity, and adjourned together to one or other of the entrances, generally to the main opening behind the big mound. There, unseen, they could watch the rooks sail slowly overhead, and the pigeons, with a sharp hiss of swiftly beating wings, drop down into the trees, and flutter, cooing loudly, from bough to bough before they fell asleep. [Pg 345] Then, after a twilight romp in and about the mouth of the burrow, the badgers took up the business of the night, and wandered away over the countryside in search of food, sometimes extending their journeys even as far as the garden of a cottage five miles distant, where Brock distinguished himself by overturning a hive and devouring every particle of a new honeycomb found therein.

Autumn, beautiful with pearly mists and red and golden leaves, again succeeded summer, and the woods resounded with the music of the huntsman's horn, as the hounds "harked forward" on the scent of fleeing fox-cubs, that had never heard, till then, the cries of the pursuing pack.

One morning, Brock lay out in the undergrowth, though the sun was high and the rest of his family slept safely in the burrow. At the time, his temper was not particularly sweet, for, on returning to the "set" an hour before dawn, he had quarrelled with his sire. Among the dead leaves and hay strewn on the floor of the chamber usually inhabited by the badgers in warm weather, was an old bone, discovered by Brock in the woods, and carried home as a plaything. For this bone Brock [Pg 346] had conceived a violent affection, almost like that of a child for a limbless and much disfigured doll. He would lie outstretched on his bed, for an hour at a time, with his toy between his fore-feet, vainly sucking the broken end for marrow, or sharpening his teeth by gnawing the juiceless knob, with perfect contentment written on every line of his long, solemn face. If disturbed, he would take the bone to the winter "oven" below, and there, alone, would toss it from corner to corner and pounce on it with glee, or, with a sudden change of manner, would grasp it in his fore-paws, roll on his back, and scratch, and bite, and kick it, till, tired of the fun, he dropped asleep beside his plaything; while overhead, the rabbits and the voles, at a loss to imagine what was happening in the dark hollows of the "earth," quaked with fear, or bolted helter-skelter into the bushes beyond the mound.

When, just before the quarrel, Brock sought for his bone, as he was wont to do on returning home, he scented it in the litter beneath a spot completely overlapped on every side by some part or other of his recumbent sire. For a few moments, he was [Pg 347] nonplussed by the situation; then, desperate for his plaything, he suddenly began to dig, and, in a twinkling, was half buried in the hay and leaves; while to right and to left he scattered soil and bedding that fell like a shower over his mother and sister. Before the old dog-badger had realised the meaning of the commotion, Brock had grabbed his treasure, and, withdrawing his head from the shallow pitfall he had hurriedly fashioned, had caused his drowsy parent to roll helplessly over. This was more than a self-respecting father could possibly endure in his own home and among his own kin, so, with unexpected agility, as he turned in struggling to recover his balance, he gripped Brock by the loose skin of the neck, and held him as in a vice from which there seemed no escape. Brock, doubtless thinking that his right to the bone was being disputed, strove vigorously to get hold of his sire, but the grip of the trap-like jaws was inflexible, and kept him firmly down till his rage had expended itself, and he was cowed by his parent's prompt, easy show of tremendous power. When, at last, the old badger relinquished his hold, Brock shook [Pg 348] himself, and sulkily departed from the "set," followed to the door by his relentless chastiser. An hour before noon, Brock heard the note of a horn—sounding far distant, but really coming only from the other side of the hill—succeeded by the eager baying of a pack of fox-hounds. Then, for a while, all was silent, but soon the cries of the hounds broke out again, away beyond the farm by the river. Evidently something was amiss. Brock, though hardly, perhaps, alarmed, shifted uneasily in his retreat under the yellow bracken, and finally, almost fascinated, lay quiet, watching and listening. Presently the ferns parted; and a fox-cub appeared in full view, treading lightly, his tongue lolling out, his jaws strained far back towards his ears, and his face wearing the look of a creature of excessive cunning, though for the time frightened nearly out of his wits. The fox-cub paused an instant, turned as if to look at something in the dark thickets by the glen, climbed the mound, and, after another hasty glance, entered his home among the outer chambers of the "set." Unknown, of course, to Brock, the leading hounds were running mute on the fox-cub's [Pg 349] scent down the path by the river. They swerved, and lost the line for a moment, then, "throwing their tongues," crashed through the briars into the fern; and at once Brock was surrounded.

Luckily, he had neither been punished too severely by his sire, nor had exhausted himself in hotly resisting the chastisement. For a few seconds, however, as the hounds pressed closely in the rough-and-tumble fray, trying to tear him limb from limb, he was disconcerted. But quickly regaining his self-possession, he began to make the fight exceedingly warm for his assailants. A hound caught him by the leg; turning, he caught the aggressor by the muzzle. His strong, sharp teeth crashed through nose and lip clean to the bone, and the discomfited hound, directly one of the pack had "created a diversion," made off at full speed, running "heel," and howling at the top of his voice. One after another, Brock served two couples thus, till the wood was filled with a mournful chorus altogether different from the usual music of the hounds.

Little hurt, except for a bruise or two on [Pg 350] his loose, rough hide, and feeling almost as fresh as when the attack began, Brock, with his face to the few foes still remaining to threaten him hoarsely from a safe distance, retired with dignity to the mound, and disappeared in the tunnel just as reinforcements of the enemy hastened up the slope.

Henceforth, even in leafy summer, he seldom remained outside his dwelling during the day, and any fresh sign of a dog in the neighbourhood of his immediate haunt never failed to fill him with rage and apprehension.

Since the time when their silvery-grey coats had turned to brownish-yellow, the badger cubs had become more and more independent of their parents; and before long, familiar with the forest paths, they often wandered alone. Yet so regular was their habit of returning home during the hour preceding dawn, that, unless something untoward happened, the last badger to reach the "earth" was rarely more than a few minutes after the first. Towards the end of autumn, however, the female cub seemed to have lost this habit; on several occasions dawn was breaking when she sought [Pg 351] her couch; and one morning she was missing from the family. Her regular home-coming had given place to meeting, in a copse over the hill, a young male badger reared among the rocks of a glen up-stream; and by him she had at last been led away to a home, which, after inspecting several other likely places, he had made by enlarging a rabbit burrow in a long disused quarry.

Brock was in no hurry to find himself a spouse; he waited till the end of winter. Meanwhile, the colour of his coat changed from yellow to full, dark grey, and simultaneously a change became apparent in his disposition. Wild fancies seized him; from dusk to dawn he wandered with clumsy gait over the countryside, little heeding how noisily he lumbered through the undergrowth. The gaunt jackhare, that, crying out in the night, hurried past him, was not a whit more crazy.

At one time, Brock met a young male badger in the furze, attacked him vigorously, and left him more dead than alive. At another time, he even turned his rage against his sire. The old badger was by no [Pg 352] means unwilling to resent provocation: he, too, felt the hot, quick blood of spring in his veins. The fight was fierce and long—no other wild animal in Britain can inflict or endure such punishment as the badger—and it ended in victory for Brock. His size and strength were greater than his father's; he also had the advantage of youth and self-confidence; but till its close the struggle was almost equal, for the obstinate resistance of the experienced old sire was indeed hard to overcome. Brock forced him at last from the corner where he stood with his head to the wall, and hustled him out of doors. Then the victor hastened to the brook to quench his thirst, and, returning to the "set," sought to sleep off the effects of the fight. When he awoke, he found that the mother badger had gone to join her evicted mate. The inseparable couple prepared a disused part of the "set" for future habitation; there they collected a heap of dry bedding, and, free from further interruption, were soon engaged with the care of a second family.

For nearly a week after his big battle, Brock felt stiff and sore, and altogether too [Pg 353] ill to extend his nightly rambles further than the boundaries of the wood. But with renewed health his restlessness returned, and he wandered hither and thither in search of a mate to share his dwelling. A knight-errant among badgers, he sought adventure for the sake of a lady-love whose face he had not even seen.

Sometimes, to make his journeys shorter than if the usual trails from wood to wood had been followed, he used the roads and byways leading past the farmsteads, and risked encounter with the watchful sheep-dogs. For this indiscretion, he almost paid the penalty of his life. Crossing a moonlit field on the edge of a covert, he saw a flock of sheep break from the hurdles of a fold near the distant hedge, and run panic-stricken straight towards him. Long before he had time to regain the cover, they swept by, separating into two groups as they came where he stood. Immediately afterwards, he saw that one of the sheep was lying on her back, struggling frantically, while a big, white-ruffed collie worried her to death. The dog was so engrossed with his victim that the badger remained [Pg 354] unnoticed. Having killed the sheep, the dog sat by, panting because of his exertions, and licking the blood from his lips. Suddenly, raising his head, he listened intently, his ears turned in the direction of the fold. Then, growling savagely, he slunk away, with his tail between his legs, and disappeared within the wood.

He had scarcely gone from sight, when the farmer and his boy climbed over the hedge near the field and hastened across the pasture. They saw the sheep lying dead, and, not far from the spot, the badger lumbering off to the covert. Instantly believing that Brock was the cause of their trouble, they called excitedly for help from the farm, and dashed in pursuit. As Brock gained the gap by the wood, he felt a sharp, stinging blow on his ribs. On the other side of the hedge, he reached an opening in the furze, and the sticks and stones aimed at him by his pursuers, as he turned downwards through the wood, fell harmlessly against the trees and bushes. The noise he made when crashing through the thickets was, however, such a guide to his movements, that he failed to baffle the [Pg 355] chase till he reached a well worn trail through the open glades. Luckily for him, as he emerged from cover a cloud obscured the moon, and he was able to make good his escape by crossing a deep dingle to the lonely fields along his homeward route, where, in the shadows of the hedges, though now the moon again was bright, he could not easily be seen.

It was fortunate for the badger, not only that the moon was hidden by a cloud as he crossed the dingle when fleeing from the wood, but also that his home was distant from the scene of the tragedy in the upland pasture near the farm. A hue-and-cry was raised, and for days the farmer's boy searched the wood around the spot where Brock had disappeared, hoping there to find the earth-pig's home. Other sheep were mysteriously killed on farms still further from the badger's "earth"; then watchers, armed with guns, lay out among the cold, damp fields to guard the sleeping flocks; and the collie, a beautiful creature whose character had hitherto been held above reproach, was shot almost in the act of closing on a sheep he had already [Pg 356] wounded, close to the corner of a field where a shepherd lay in hiding.

The farmer and his boy were chaffed so unmercifully—for this story of the badger was now considered a myth—that they grew to hate the very name of "earth-pig," and to believe that after all they must have chased through the wood some incarnation of Satan.

V.

HILLSIDE TRAILS.

Several times during his search for a mate, Brock struck the trail of a female badger, and followed its windings through the thickets and away across the open fields towards the the distant valley, only, however, to lose it near some swollen brook or on some well trodden sheep-path. The female had evidently come to a little copse on the crest of a rugged hill overlooking the river, and, after skirting a pond where wild duck sheltered among the flags, had retraced her steps. Brock's most frequented tracks led close to the spot where the stranger's return trail joined the other near an opening from an almost impenetrable gorse-cover into a marshy fallow. There, late one night, he found, as he crossed the opening, that the female badger had travelled forward, but had not yet returned. Revisiting the spot some minutes afterwards, he discovered that the backward "drag" was strong on the damp grass. He followed it quickly, and, in a stubble beyond the gorse, came up at last with the object of his oft-disappointed quest. She was a widow badger, older and more experienced than Brock, but smaller and of lighter build.

Perhaps because she wished to test the loyalty of her new lover, and to find whether he would fight for her possession with any intruder, she resisted his advances, and refused to go with him to his home. So he followed her far away to her own snug dwelling on the fringe of the moorlands. Thence, with the first streak of dawn in the south-eastern sky, he hurried back to his lair.

Early next evening, Brock went forth to meet his lady-love; and throughout the long night and for nights afterwards he wandered at her side, till, concluding that no other suitor was likely to appear, she accompanied him to his home, and entered on the season's house-keeping in the central chamber of the great "set" where he had been born. There they lived happily, and without the slightest annoyance from the old badgers; and, since the time of the spring "running" was over, they wandered no further afield than in the cold winter nights. Filled with the joy of the life-giving season, they often romped together in the twilight for half an hour at a time, chasing one another in and out of the entrances to the "set," or kicking up the soil as if they suddenly recollected that their claws needed to be filed and sharpened, or standing on their hind-feet and rubbing their cheeks delightedly against a favourite tree—grunting loudly in their fun the while, and in general behaving like droll, ungainly little pigs just escaped from a stye. At last, their frolic being ended, they "bumped" away into the bushes, and, meeting on the trail beyond, proceeded soberly towards the outskirts of the wood.

As in the previous spring, the big burrow was soon the scene of family affairs other than those of the badgers. By the end of February, there were cubs in the vixen's den, and both the wood-mice and the rabbits were diligently preparing for important family events. Brock's companion, unlike himself [Pg 360] was not accustomed to a house inhabited by other tenants. None but members of her own family had dwelt in the "earth" near the moor; and, being somewhat exclusive in her ideas, she strongly resented the presence of the vixen in any quarter of her new abode. A little spiteful in her disposition, she lurked about the passages, and by the mound outside the entrance, intending to give her neighbour "a bit of her mind" at the first opportunity. But since she did not for the present care to enter the vixen's den, that opportunity never came till her own family arrangements claimed her undivided attention, and effectually prevented her from following the course of action she had planned.

In the first week of April, the badger's spring-cleaning began in downright earnest. The old bedding of fern, and hay, and leaves was cleared entirely from the winter "oven," and, after a few windy but rainless days and nights, when the refuse of Nature's woodland garden was dry, new materials for a cosy couch were carried to the lair, and arranged on the floor of the roomy chamber where Brock's mother had brought him [Pg 361] into the world. The badgers' methods of conveying the required litter were quaintly characteristic, for the animals possessed the power of moving backward almost as easily and quickly as forward. They collected a pile of leaves, and, grasping it between their fore-legs, made their way, tail first, to the mound, and thence, in the same manner, along their underground galleries, as far as the place intended for its reception, strewing everywhere in the path proofs of their presence, quite sufficient for any naturalist visiting their haunts.

On a dark, wet night rather less than a fortnight after they had completed their preparations, when Brock returned to his home for shelter from the driving storm, three little cubs were lying by their mother's side.

The training of the badger-cubs during the first two months was left wholly to their dam; but afterwards Brock shared the work with his mate, teaching the youngsters, by his example, how to procure food, and, at the same time, to detect and to avoid all kinds of danger. In so doing, he simply acted towards his cubs [Pg 362] as his sire had acted towards him. Apart from family ties, however, his life—that of a strong, deliberate animal, self-possessed in peril and in conflict, yet shy and cautious to a fault—was of extreme interest to both naturalist and sportsman.

Five young foxes, as well as the vixen, now dwelt in the antechamber near the main entrance of the "set," and the presence of this numerous family became, for several reasons, so objectionable to the she-badger, that, about the middle of May, the antipathy which, since her partnership with Brock, she had always felt towards the vixen, was united with a fixed determination to get rid of her neighbours. She was too discreet, however, to attempt to rout them during the day, when some dreaded human being might be attracted by the noise; so she endeavoured to surprise the vixen and her cubs together at night.

For a while, she was unsuccessful. She happened to frighten them by an impetuous, blustering attack in the rear, from which they easily escaped; thus her difficulties had been increased, since the objects of her aversion became loath to stay in the "earth [Pg 363]" after nightfall. But at last, probably more through accident than set purpose, the badger out-manœuvred the wily foxes.

Lying one evening in the doorway, she heard the vixen, followed by the young foxes, creeping stealthily from the den. Retreating quickly, she barred their exit, thus compelling them to return to their lair; then she took up her position in the neck of the passage, and waited patiently till midnight before commencing her assault. At last, in the dense darkness, she crawled along the winding tunnel, and, directly, the den was the scene of wild confusion and uproar, as its inmates leaped and tumbled over each other in their frantic efforts to escape. For a few minutes, the advent of danger unnerved them; then, as if peculiarly fascinated by the grim, motionless enemy blocking their only outlet, they began an aimless, shuffling dance, baring their teeth and hissing as they lurched from side to side. Their suspense was soon ended. The badger, emerging partly from the passage, gripped one of the cubs by a hind-leg, and dragged it backwards along the passage to the thicket outside, where, after worrying her victim unmercifully, she ended its life by [Pg 364] crushing its skull, above the muzzle, into fragments between her teeth.

Once more, but this time furious with the taste of blood, she hurried to the den; and the scene of fear and violence was repeated. Her third visit was futile: the vixen with the other cubs had bolted into the main gallery, and escaped thence to the wood, through an old opening, almost choked with withered leaves, at the back of the "set."

They never returned, but the following spring a strange vixen from the rocks across the valley came to the burrow, gave birth to her young, and, in due course, without loss, was evicted by Brock's relentless mate.

"HE CLIMBED FROM HIS DOORWAY, AND STOOD MOTIONLESS, WITH UPLIFTED NOSTRILS, INHALING EACH BREATH OF SCENT." To List

On the night after the death of the fox-cubs, when Brock was led by the she-badger to the spot where her victims lay, he noticed that man's foot-scent was strong on the grass around, and also that his hand-scent lingered on the fur of the slain animal. Often, during the succeeding two months, he was awakened in the day by quick, irregular footsteps overhead; and later, when he climbed from his doorway, and stood motionless, with uplifted nostrils, inhaling each breath of scent, he found that the dreaded signs of man were [Pg 365] numerous on the trail, on the near beech-trunk, and even on the mound before the "set." Once, on returning home with his family, he was greatly alarmed to discover that in the night the man had visited his haunts, and that a dog had passed down the galleries and disturbed the bed on which he slept. Henceforward, he used the main opening as an exit only, and invariably entered the "set" by the opening through which the vixen had escaped from his mate, passing, on his way, the mouth of a side-gallery connected with the apartments occupied by his old sire and dam, together with their present family. Eventually, through these precautions, he saved his principal earthworks from destruction.

Had Brock been able to ascertain the meaning of man's frequent visits to the neighbourhood of his dwelling, he would have sorely lamented the killing of the young foxes by the female badger. In the eyes of the Hunt, vulpicide was an unpardonable crime, whether committed by man or beast; and, when the dead fox-cubs were shown to the huntsman, he vowed vengeance on the slayer. Because of a recent exchange, between the [Pg 366] two local Hunts, of certain outlying farms, it happened that this huntsman was not he who in past seasons had tethered his horse near the "set" while he "drew" the cover on foot. The new-comer soon discovered the "earth"; but after a brief examination, from which he concluded, because of the strong taint still lingering, that it was tenanted by a fox, he walked away towards the farm. Fearing a reprimand from the Master if the mysterious slaughter of the foxes could not be explained, he made careful enquiries of the farmers, by whom he was told of the badger and the sheep, as well as of the poacher who had seen Brock's sire in the upland fields two years ago; but he laughed at the first tale, and for want of adequate information paid no heed to the second. Nevertheless, when he again visited the "earth," and, stooping, saw the withered leaves and fern, and detected, not now the scent of a fox, but the scent of half a dozen badgers, his sluggish brain began to move in the right direction.

Stories he had heard by the lodge fireside when he was a lad, casual remarks dropped by followers of the Hunt, questions asked him by an inquisitive [Pg 367] boy-naturalist—he slowly remembered them all; and then the revealing light dawned on his mind, that no animal but a badger could with ease have broken the limbs of a fox-cub, and cracked the skull as though it were a hazel-nut. Filled with a sense of self-importance, befitting the bearer of a momentous message, the huntsman rode away in the breathless summer twilight to the country house where the Master lived, and presently was shown into the gun-room to wait till dinner was over.

The Master prided himself on his love of every kind of sport; and before the huntsman had finished a long, rambling story of the woodland tragedy he had formed his plans for the punishment of the offender and was writing a brief, urgent letter to a distant friend. As the result, a few days afterwards three little terriers, specially trained for "drawing" a badger, arrived at the Master's house, and were accommodated in a vacant "loose-box" in the stables. Late at night, one of these was introduced to the "set," and from the experiment the Master was led to believe that, though the place, as he surmised, was empty of its usual [Pg 368] tenants at the time, it held sure promise of sport for an "off" day, as soon as the otter-hounds, now about to hunt in the rivers of the west, had departed from the neighbourhood. Meanwhile, according to his strictest orders, the little terriers were well fed, regularly exercised, and kept from quarrelling, and their coats were carefully brushed and oiled that they might be as fit as fiddles for the eventful "draw."

The Master was a rigid disciplinarian in all matters concerned with sport. His servants, one and all, from the old, white-haired family butler down to the little stable-boy, idolised him, but never presumed to disobey his slightest command. For many years before he came to live at the mansion, the Hunt had fallen into a state of extreme neglect; the pack was one of the worst in the kingdom, the subscriptions were irregular, the kennel servants were ill-paid, the poor cottagers never received payment for losses when Reynard visited their hen-coops, and even the farmers began to grumble at needless damage to their hedges, and to refuse to "walk" the puppies. But the new Master had changed all this. He bore his share, [Pg 369] but no more, of the expense caused by the reforms he at once introduced, and he reminded his proud yet stingy neighbours that the pack existed for their sport as much as for his own, that arrears were shown in his secretary's subscription-books, and that, unless the funds were augmented, he would reconsider the step he had taken in accepting the Mastership. Useless servants, useless hounds, and merely ornamental members of the Hunt, alike disappeared; and with system and discipline came season after season of prosperity, contentment, and justice, till it seemed that the best old traditions of British sport were revived in a community of hard-working, rough-riding fox-hunters, among the isolated valleys of the west.

As might be inferred from the personality of the Squire, everything was in apple-pie order on the glorious summer morning when he and his huntsmen made their way down river to the wood inhabited by Brock. A complete collection of tools—crowbar, earth-drill, shovels, picks, a woodman's axe, and a badger-tongs that had been used many years ago to unearth a badger in a distant county, and ever since had occupied a corner in [Pg 370] the Squire's harness-room—had already been conveyed to the scene of operations, together with a big basket of provisions and a cask of beer, it being one of the Squire's axioms that hard work deserved good hire. Four brawny labourers were also there; and, near by, each in leash, the three little terriers lay among the bilberries. Punctually at the time appointed, the work of the day began. A terrier was led to the main entrance of the "set," but, to the dismay of the huntsman, he refused to enter. When, however, he was brought to the entrance that artful Brock had lately used, he at once became keenly excited, dragged at his leash, and, on being freed, disappeared in the darkness of the burrow. The Master knelt to listen; and presently, as the sound of furious growls and barks came from the depths, he arose, saying: "Now, my men, we may begin with picks and shovels; our badger is at home."

What followed, from that early summer morning till twilight shadows fell over the woods, and men and dogs, completely beaten, wended their way homewards along the river-path, may best be told, perhaps, in a bare, simple narrative of events as they occurred. [Pg 371]

When the terrier went "to ground," he crawled down a steep, winding passage into a hollow, from twelve to fifteen feet below the entrance. Thence, guided by the scent of a badger, he climbed an equally steep passage, to a gallery about six feet below the surface. Following the gallery for a yard or so, he came to a spot where it was joined by a side passage, and here, as well as in the gallery beyond, the scent was strong. He chose the side passage, crept down a slight declivity, and came where Brock's sire had, a few minutes before, been lying asleep, while his mate and cubs occupied the centre of the chamber. Awakened by the approach of the terrier, the she-badger and her offspring had hurried to another chamber of the "set," and the male had retreated to a blind alley recently excavated back towards the main gallery. The terrier, keeping to the line he had struck at the sleeping place, found the male badger at work there, throwing up a barrier between himself and his pursuing enemy, and at once diverted his attention by feinting an attack in the rear. For two hours, the game little dog, avoiding each clumsy charge and yet not giving the badger [Pg 372] a moment's peace, remained close by, while the men cut further and further into the "set," till they stood in the first deep chamber through which the terrier had passed. Then the terrier came out to quench his thirst, and was led away by the huntsman to the river, while the second dog was speedily despatched to earth, that the badger might be allowed no breathing space during which he could bury himself beyond the reach of further attack. The second dog, on coming to the junction of the passage and the gallery, chose the alternative line of scent in the gallery, and wandered far away into the chamber where Brock, whose family had descended some time before to the winter "oven," awaited his coming. When the faint barking of the second terrier told that the badger had seemingly shifted his quarters to an almost incredible distance from the trench, the faces of the Squire and his assistants evinced no little surprise. For a moment, the men were inclined to believe that the dog was "marking false," but, presently, their doubts were dispelled, and their hopes revived, as the sounds indicated that the terrier, contesting hotly every inch of the [Pg 373] way, was retreating towards them before his enraged enemy. The labourers resumed work, though not with the confidence of the early morning, when their task seemed lighter than the experienced Master would admit. Hour after hour they toiled; the dogs were often changed; and at last the trench was long enough to be within a yard or so of the spot where the dog was engaged. Then, to the mortification of the sportsmen, the sounds of the conflict suggested another change: Brock was retiring leisurely to his chamber. The earth-drill was soon put into play, and the badger's position discovered, but directly afterwards the animal again moved, this time to the deep "oven" below.

Night was now rapidly closing over the woods, and the weary, disappointed men and dogs reluctantly gave up their task. The Squire admitted that on this occasion, at any rate, he was fairly and squarely beaten. Brock and his mate are still in possession of the old burrow beyond the farm; and Brock's sire, a patriarch among badgers, lives, as the comrade of another old male, among the boulders of a rugged hillside a mile from the "set."

THE HEDGEHOG.

[Pg 377]

I.

[Pg 375]
Top

A VAGABOND HUNTER.

At the lower end of our village, the valley is joined by a deep ravine through which a sequestered road—hidden by hawthorn hedges, and crossed by numerous water-courses where the hillside streams, dropping from rocks of shale, ripple towards a trout-brook feeding the main river—winds into the quiet country. The rugged sides of the ravine are thickly clothed with gorse and brambles, and dotted with hazels, willows, and oaks. This dense cover is inhabited by large numbers of rabbits; in a sheltered hollow half-way up the slope a badger has dug his "set"; and in the pastures above the thickets a fox may be seen prowling on almost any moonlit night. Past the gorge, the glen opens out in rich, level pastures and meadows bounded on either [Pg 378] side by the hills. The nearest farmsteads are built high among the sunny dingles overlooking the glen, and the corn and the root-crops are grown on the slope beyond the broad belts of gorse and bramble.

In winter, the low-lying lands are seldom visited by the peasantry, except when the dairymaid drives the cattle to and fro, or the hedger trims the undergrowth along the ditches. Though the sportsman with gun and spaniels and the huntsman with horse and hounds are frequently heard in the thickets, they never visit the "bottom," unless the partridges fly down from the stubble, or the hare, pursued by the beagles, takes a straight line from the far side of the glen to a sheep-path leading up the gorge. And in summer, except when the fisherman wanders by the brook, and the haymakers are busy in the grass, the glen is an undisturbed sanctuary, given over to Nature's wildlings, where, in safety, as far as man is concerned, they tend their hidden young.

In this quiet, windless place, on the day when first the haymakers came to the meadows, five little hedgehogs were born in a nest among the roots of a tree, deep [Pg 379] in the undergrowth of a tangled hedgerow. The nest was made of dry grass and leaves, and with an entrance so arranged amid the "trash," that, when the parent hedgehogs went to or from their home, they pushed their way through a heap of dead herbage, which, falling behind them, hid the passage from inquisitive eyes.

It may be asked why such a warm retreat was necessary, inasmuch as the hedgehog sucklings came into the world in the hottest time of the year. Nature's reasons were, however, all-sufficient; the little creatures, feeble and blind, needed a secure hiding place, screened from the changeful wind of night and from every roving enemy. The haymakers, moving to and fro amid the swathes, knew nothing of the hedgehogs' whereabouts; but when the dews of night lay thick on the strewn wild flowers, the parent "urchins," leaving their helpless charges asleep within their nest, wondered greatly, while they hunted for snails and slugs in the ditch, at the quick change that had passed over the silent field.

For a week or more, the spines sprouting from round projections on the bodies of the [Pg 380] young hedgehogs were colourless and blunt, and so flexible that they could have offered no defence against the teeth or the claws of an enemy; while every muscle was so soft and feeble that not one of the little animals was as yet able to roll itself into the shape of a ball. The spines, however, served a useful purpose: they kept the tender skin beneath from being irritated by the chance touch of the mother hedgehog's obtrusive quills.

Soon the baby hedgehogs' eyes opened wide to the pale light filtering between the leaves at the entrance to the chamber, and their spines, gradually stiffening, assumed a dull grey colour. Then, one still, dark night, the little creatures, with great misgiving, followed their parents from the nest, and wandered for a short distance beside the tangled hedge. Presently, made tired and sleepy and hungry by exercise and fresh air, they were led back to their secret retreat, where, after being tended for a few moments by their careful mother, they fell asleep, while their parents searched diligently for food in the dense grass-clumps left by the harvesters amid the briars and the furze.

Henceforth, every night, they ventured, [Pg 381] under their mother's care, to roam afield, their journeys becoming longer and still longer as their strength increased, till, familiar with the hedge-row paths, they were ready and eager to learn the rudiments of such field-craft as concerned their unpretending lives.

A glorious summer, far brighter than is usual among the rainy hills of the west, brooded over the countryside. The days were calm and sunny, but with the coming of evening occasional mists drifted along the dingles and scattered pearl-drops on the after-math; and the nights were warm and starlit, filled with the silence of the wilderness, which only Nature's children break. The "calling season" for the hare had long since passed, and for the fox it had not yet arrived; so the voices of the two greatest wanderers on the countryside were not at this time heard.

A doe hare had made her "form" hardly twenty yards from the hedgehogs' nest, and night after night, just when the "urchins" moved down the hedge from the old tree-root, she ambled by on her way to the clover-field above the heath. [Pg 382]

Once, a little before dawn, a fox, coming to drink at the brook, detected the scent of the hedgehogs near a molehill, followed it to the litter of leaves by the tree, and caused considerable alarm by making a vigorous attempt to dig out the nest; but, probably because of the dampness of the loamy soil, he failed to determine the exact whereabouts of the hedgehog family; and, after breaking a tooth in his vain efforts to cut through a tough, close-fibred root, he made his way along the hedge, and soon disappeared over the crest of the moonlit hill. But the next night, when the wind blew strong, and the rain pattered loudly on the leafy trees, he came again to the "urchins'" haunt. The doe hare had long since rustled by, and the hedgehogs were busy munching a cluster of earthworms discovered in a heap of refuse not far from the gate, when Reynard stole over the fence-bank, and sniffed at the nest. Not finding the family at home, he followed their scent through the ditch, and soon surprised them. To kill one of the tiny "urchins" was the work of a moment; then, made eager by the taste of blood, the [Pg 383] fox turned on the mother hedgehog and tried to fix his fangs in the soft flesh beneath the armour of her spines. But, feeling at once his warm breath, she, with a quick contraction of the muscles, rolled herself into a prickly ball, and remained proof against his every artifice. He was a young fox, not yet learned in the wiles of Nature's feebler folk, and so, when he had recovered from his astonishment, he pounced on the rigid creature, and, thoughtlessly exerting all his strength, endeavoured to rend her in pieces with his powerful jaws. He paid dearly for his temerity. The prickly ball rolled over, under the pressure of his fore-paws, the sharp points of the spines entered the bare flesh behind his pads, and as, almost falling to the ground, he bit savagely to right and left in the fit of anger which now possessed him, his mouth and nostrils dripped blood from a dozen irritating wounds. Thoroughly discomfited, he leaped back into the field, where, sick with pain, he endeavoured to gain relief by rubbing his muzzle vigorously in the grass and against his aching limbs. Then, sneezing violently, and with his mouth encrusted [Pg 384] with froth and loam, he bolted from the scene of his unpleasant adventure, never pausing till he reached his "earth" on the hillside, in which, hidden from the mocking gaze of other prowlers of the night, he could leisurely salve his wounds with the moisture of his soft, warm tongue, and ponder over the lessons of his recent experience.

By far the most intelligent and powerful enemy of the young hedgehogs was the farmer's dog; but, as he slept in the barn at night, and generally accompanied the labourers to the upland fields by day, they escaped, for a while, his unwelcome attentions. Foes hardly less dreaded, because of their insatiable thirst for blood, were two polecats living in a hole half-way up the wall of a ruined cottage not far from the hillside farm-house. The polecats, however, were so occupied with the care of a family, that, finding young rabbits plentiful in the burrows on the heath, they seldom wandered into the open fields, till the little "urchins," ready, at the first sign of danger, to curl themselves within the proof-armour of their growing spines, were well able to resist attack. [Pg 385]

The hedgehogs were about three months old, and summer, brief and beautiful, was passing away, when an incident occurred that might have proved disastrous, though, fortunately, it resulted only in a practical joke, such as Nature often plays on the children of the wilds. One calm, dark night, while they were busy in the grass, a brown owl, hunting for mice, sailed slowly by. Now, the brown owl, in spite of proverbial wisdom gained during a long life in the dim seclusion of the woods, is occasionally apt to blunder. Her character, indeed, seems full of quaint contradictions. As she floats through the moonlight and the shadows of the beech-aisles of Dollan, she appears to be a large bird, with a philosophic contentment of mind—an ancient creature that, shunning the fellow-ships of the garish modern day and loving the leisure and the solitude of night, dreams of the past. But, beneath its loose feathery garments, her body, hardly larger than that of a ringdove, is altogether out of proportion to her long, narrow head and wide-spreading talons. Visions of the past may come to her, as, blinking at the light [Pg 386] of day, she sits in the hollow of the tree, but at night she is far too wide-awake to dream. And so great are the owl's powers of sight and hearing, and so swift is her "stoop" from the sky to the ground, that the bank-vole has little chance of escape should a single grass-stalk rustle underfoot when she is hovering near his haunt. Far from being shy and retiring in her disposition, the brown owl, directly night steals over the woodlands, is so fearless that probably no animal smaller than the hare can in safety roam abroad.

As the owl flew slowly past the fence, she heard the faint sound of a crackling shell—the hedgehogs were feeding on snails. She could barely distinguish a moving form in a tangle of briars, but its position discouraged attack; so she flew away and continued to hunt for mice. Presently, returning to the spot, the owl was once more attracted by the sound of some creature feeding in the grass; and, detecting a slight movement beside the briars, she swooped towards the ditch, grasped one of the "urchins" in her claws, and rose into the air. Her quarry, feeling the sudden grip of the sharp talons, made a desperate, convulsive [Pg 387] movement, and the owl found, to her astonishment, that her grasp had shifted, and that she was holding, apparently, a hard bunch of thorns. Nevertheless, she tightened her grasp; but an unendurable twitch of pain, as the spines entered her flesh immediately above the scales of her talons, caused her to drop the hedgehog into the leaf-mould of the ditch. Immediately afterwards, she herself, eager to find out the cause of her discomfiture, dropped also to the earth, and, standing beside the hedgehog, clawed savagely at the motionless creature, seeking some defenceless point among the bristling spines. At last, her patience exhausted, the owl gave up the ineffectual assault, and glided away into the gloomy night. Unhurt, but for a slight wound inflicted when first the bird descended, the hedgehog crawled back to the brambles, where the rest of her family were still busy with the snails, and joined them in their feast.

Autumn's sere leaves had fallen from the trees, and the hedgehogs had found such a plentiful supply of all kinds of food that they were ready for their winter sleep, when a gipsy boy, the proud possessor of a [Pg 388] terrier trained for hunting hedgehogs, set forth in haste one evening from his tent by the wayside above the farm. The boy was smarting from cruel blows inflicted by his drunken parents, who, after unusual success in disposing of baskets and clothes-pegs, had spent much of the day's profit in a carouse at the village inn. Having escaped a continuance of his parents' brutalities, and eluded their ill-conducted pursuit, the young gipsy, in the company of his only friend, soon forgot his miseries as his thoughts turned to a vagabond's rough sport in the stillness of the harvest night. Thrusting a long stick here and there into the briars, he strolled along by the fence, till his dog, diligently beating in line amid the undergrowth, gave a quick yelp of delight, and, an instant later, a curled-up hedgehog rolled down into the ditch. The boy placed the animal in his ragged handkerchief, the corners of which he was proceeding to tie together when the terrier again attracted attention with unmistakable signs of a "find." For a few brief minutes sport was keenly exciting, but at last all the "urchin" family, with the exception of one member, were captured, and the boy, now thoroughly happy, [Pg 389] his pockets and handkerchief heavy with spoil, turned homewards through the darkness. Next morning, the slain hedgehogs, baked in clay among the hot ashes of a fire of rotten twigs, formed the principal item in the gipsies' bill of fare, and the terrier enjoyed the remnants of the meal.

The hedgehog surviving the gipsy's raid was a young female, that, while the terrier beat the fence, remained quietly munching a large lob-worm at the foot of a mound a dozen yards away, and so knew nothing of the fate of her kindred.

The last weeks of the year passed uneventfully, as far as her little life was concerned; then, as the nights grew longer and the cold increased, she set about preparing in earnest for her long, deep sleep.

In a sheltered spot close to the woodlands, where, a month before, a badger had unearthed a wild bee's nest, she collected a heap of withered oak-leaves, hay, and moss, and with these simple materials made a large, snug nest, a winter house so constructed that the rain might trickle down to the absorbent soil beneath. For [Pg 390] a little while, however, she did not enter into her unbroken rest. Still, nightly, she roamed abroad, moving in and out of the dried herbage everywhere strewn in her paths among the tree-roots, till the sapless leaves impaled on the sharp points of her spines formed such a cluster that she lost all semblance of a living creature. Insects were becoming rarer and still rarer as the year drew to its close, and those surviving the frosts retired to countless secret chambers at the roots of the moss and under the tough bark of the trees. The lizards sought shelter in warm hollows deep below the piles of stones left here and there by the labourers, when, every spring, they cleared the freshening fields. And the big round snails, the luscious tit-bits of the hedgehog's provender, crept into the holes of the red mice and into the chinks of walls and banks, where, protected by their shells, each being fastened to its resting place by a neat rim of hardened glue, they lived unconscious of decay and gloom. Then the hedgehog, having become drowsier and still drowsier with privation and cold, ceased to wander from [Pg 391] her nest at dark, and began that slumber which was to last till the sweet, warm breath of spring awoke her, and other wildlings of the night, to a life among the early primroses and violets.

II.

[Pg 392]
Top

AN EXPERIENCE IN SNAKE-KILLING.

The many changes of winter passed over the countryside; tempests raged, rain beat down in slanting sheets or enveloped the fields in mist, snow fell heavily and then vanished before the breath of a westerly breeze, black frost held the fields for days in an iron clutch, and sometimes, from late dawn to early dusk, the sun shone clearly in the southern sky. The sportsman with his spaniels wandered by the hedge, the huntsman with his beagles chased the hare across the sodden meadows, and the report of a gun or the note of a horn echoed among the surrounding hills. But in spite of changing weather and dangers from unresting foes, the hedgehog slept peacefully within her nest of withered [Pg 393] leaves till awakened by the whisper of the warm south-western wind.

It was a calm day towards the end of March when the hedgehog awoke. Gradually, since the winter solstice, the shadows of noon, cast from the wooded slope across the meadows in the glen, had become shorter; and now, when the sun reached its meridian, its beams fell directly on the spot where the hedgehog rested among the littered leaves. She felt the strange and subtle influence of spring, and crawled feebly from her retreat. The light above her nest was far too brilliant for her eyes, which had been closed for three long months, and were at best only accustomed to the gloom of night, so she sought the shadow of a tree-trunk near, and there, for a while, remained quite motionless. With the leaves of last autumn still clinging thickly to her spines, she seemed an oddly fashioned creature belonging to a distant age, a little Rip Van Winkle of the woods, with a new, quick world of unfamiliar joys and sorrows claiming her half-conscious life. Extremely feeble from cold and privation, [Pg 394] and knowing, as all Nature's wildlings seem to know, that sunlight brings with it health and strength, she presently left the shadow of the tree-trunk, and, closing her eyes, basked in complete enjoyment of the balmy day. The heat and the gentle wind soon dried her armour of spines and surcoat of leaves. Stealing in through the tunnel left open when the hedgehog came forth from her sleep, the wind cleansed and ventilated the nest, and soon all traces of winter's mustiness had vanished from both herself and her home. By sundown, the "urchin" had gained strength that enabled her to wander slowly into the meadow, where she found sufficient food to stay her growing hunger.

During the first few nights, her appetite, though keen, was easily satisfied, for the digestive organs, unaccustomed to their work, could not retain much nutriment, and hours of slumber seemed necessary after every trifling meal. But gradually her powers were restored, till almost any kind of fresh animal matter that came in her way was greedily devoured. A spider sleeping in a folded leaf, a fly hiding beneath a stone, a snail, a slug, [Pg 395] a worm, a frog, a weakling bird fallen from an early nest, a lizard, or a snake—all alike were welcome as she thrust her damp, blunt snout, that looked like a little fold of black rubber, here and there amid the herbage.

Her eyesight was faulty—she had no great need of it; her enemies were few, and she did not live the life of the hunted that fear each footfall on the grass; but, as if to balance all deficiencies, her sense of smell was singularly acute, so that she could follow with ease the trail of a beetle or of an earthworm in its windings over the soil. The eggs and young of the lark, the corncrake, the partridge, or of any other bird that built on the ground, were never safe once the hedgehog had crossed the lines of scent left by the parents around their nest. Even the robin and the wren, nesting in holes along the hedge, and the field-mouse in its chamber sheltered by the moss, were at any time likely to have their family affairs most cruelly upset. The wild-bee's sting could not save her honeyed cells and helpless grubs, and the sharp-fanged adder, writhing from the hedgehog's sudden [Pg 396] bite, would hurl itself in vain against the prickly ball that instantly confronted each counter attack.

The hedgehog's first experience of snake-killing occurred late one evening, when she discovered a viper, some distance from its hole, coiled asleep on a bare patch of soil where the sunlight had lingered at the close of day. Her manner instantly changed; she became eager and alert. Pausing only a second to make sure of her attack, she bit the snake sharply near the neck, then, withdrawing her head and limbs into the shelter of her spines, rolled over, an inanimate ball. The viper, mad with pain, thrust back its head from its sinuous coils, rose, and struck with open jaws at its assailant. Its fangs closed strongly, but failed to get a grip, and the smooth underside of its throat glanced past the hedgehog's slanting prickles with such force that the whole body of the snake was lifted from the ground, and fell, like a bent arrow, about a yard behind its foe. Again the snake rose, and struck with no effect; but this time the stroke, coming from the rear, was met by the sharp points of the spines, and the adder's mouth dropped blood [Pg 397] from a clean-cut wound on the upper edge of the palate. Repeatedly, the snake, hissing loudly and fighting for its life, attacked its armoured enemy—at first dashing itself senselessly against the sharp points of the hedgehog's spines, then, with caution, swaying to and fro its bleeding head and snapping harmlessly at an apparently unguarded spot, till, from sheer exhaustion and pain, and with its store of poison almost exhausted, it retired from the unequal combat and slowly wriggled into the grass. Presently, the "urchin" uncoiled, and, as soon as the inquisitive little snout discovered the whereabouts of the snake, started in pursuit. With a hard, firm bite, she luckily managed to break the backbone of the viper; then, at once, she again assumed the shape of a ball. Desperate now, the snake expended all its remaining strength in wild attacks, till, limp and helpless, and utterly at the mercy of the hedgehog, it lay outstretched. Then the relentless hedgehog, assured that her prey was quite defenceless, severed almost every bone in its body, tore the scales from the flesh, and fed to repletion.

Such a struggle often happens in the [Pg 398] fields and the woodlands. During the first few weeks of life, the hedgehog, if its parents are absent, may be at the adder's mercy; but, later, the tables are completely turned, the once helpless creature becomes the strong aggressor, and is revenged by removing, not only an enemy, but a rival subsisting on food often similar to that which is its own.

For a while after her awakening, the hedgehog fed chiefly on the big earthworms which, induced by the increasing warmth, forsook the deep recesses of their burrows, and tunnelled immediately beneath the grass-roots, coming forth at night to lie outstretched amid the undergrowth. She had, of necessity, to match their fear by her excessive cunning. They frequently detected her presence by the slight vibrations of the soil beneath her soft, slow-moving feet, and hurriedly withdrew from her path, but more often she surprised and captured them by the simple artifice of waiting and watching beside the burrows where scent was fresh, and where, notwithstanding the noises reaching her from above, she could readily distinguish the sounds of stretching, gliding [Pg 399] bodies moving to the surface through the tortuous passages below.

She soon became a wanderer, deserting her winter nest, and roaming nightly further and yet further from the valley meadows, till she reached a rough pasture at the end of the glen. In a thick hedgerow skirting a secluded pond among alders and willows, she found food unexpectedly varied and plentiful. Luscious snails, with striped yellow and brown shells, were so common in the ditch beyond a certain cattle-path, that, even after a whole day's fast, her hunger was quickly appeased.

April drew near, the leaves of the trees expanded, and the voice of the night wind in the branches changed from a moan to a whisper. At noon, flies came forth to bask on the stones; the furze, decked with yellow flowers, was visited by countless bees; and bronze-winged beetles crept among the thorny branches of the hawthorn and the sloe. The hedgehog knew little of the pulsing life of mid-day, but at dusk she sometimes found a tired fly, or bee, or beetle, hiding in the matted grass [Pg 400] beneath the gorse, and so was made aware of summer's near approach.

Among the flags and the rushes of the pond, a pair of fussy moorhens built their nest on an islet of decayed vegetation clustered round a stone. At all hours of the day, the birds sailed gaily hither and thither, or wandered, happy and impulsive, along the margin of the pool. No care had they, and the solitude of their retreat seemed likely never to be disturbed, till, one moonlit night, the fox, that last year had killed the baby hedgehog in the glen, stole through the shadows of the alders, caught the scent of the moorhens, and approached the nest where the female was brooding over her eggs. The bird had watched the fox's movements since first he appeared on the bank beyond the trees. Quietly she dropped into the pond beside the nest, dived, came up on the far side of the islet, and stayed there, with only her head above the surface of the water. She saw, with fear, the fox approach her nest, and recognised that it was hardly possible for her treasures to be saved, when, suddenly, her mate, having doubtless [Pg 401] watched the marauder as closely as she herself had done, walked out of a reed-clump two or three yards from her hiding place, and, in full view of the fox, swam slowly to and fro, beating his wings as if in mortal pain. Without the slightest hesitation, Reynard, thinking to obtain an easy prize, plunged into the pond, but the bird just managed to elude him, and to flutter into another reed-clump a short distance away. Completely deceived by the ruse, the fox was drawn further and further from the nest, till he reached a distant corner of the pond, when, to his astonishment, the moorhen vanished, leaving him to a vain search which at last so much annoyed him that, instead of returning along the bank towards the nest, he crossed the glen, trotted up the cattle-path, and entered the dense thicket on the slope.

With most wild creatures, fear seems to be a feeling that quickly comes and quickly goes. But over some of Nature's weaklings, fear seems to throw a spell that remains long after the danger has passed; as, for instance, in the case of a rabbit hunted by a stoat, or of a vole pursued by [Pg 402] a weasel. The animal trembles with fright, cries as if in pain, and limps, half-paralysed, towards its home, some time after its pursuer may have turned aside to follow a line of scent leading in a quite opposite direction. Now and then, a young rabbit is so overcome by fright, that the sly, watchful carrion crow obtains, with little trouble, an unexpected meal. The birds of the hedgerow—finches, robins, and the like—are also subject to the distressing influence of fear, directly they catch sight of a hungry weasel "performing" in the ditch. When the weasel sets itself to lure any such creatures, its movements are remarkably similar to the contortions of a snake; and the birds, fascinated as their enemy's strange actions are rapidly repeated, flutter helplessly from spray to spray, till one or other becomes a victim and the weasel ambles off with its prey. Then, released from the spell, the birds proceed to mob the bloodthirsty tyrant, and, at times, with such effect that he is compelled, before making good his escape, to resort to stratagems similar to those that previously held the birds enthralled. Reynard seems to have learned from the [Pg 403] weasel's manœuvres, for he, too, is wont to entice the rabbits towards him by extraordinary methods, twirling round, like a cat, in pursuit of his tail, and affording such a spectacle to any onlookers that they must needs, from sheer curiosity, find out the meaning of a woodland farce, which, alas! is often followed by a tragedy. It is not known that the fox ever succeeds in fascinating the moorhen; the bird, directly she caught sight of his circling form, would probably dive, and in the cool refuge of the water, her sharp eyes peeping from between the flags, would wisely conclude that such an unaccountable display meant danger. It is, however, tolerably certain that the influence of fear seldom causes a nesting bird, or a breeding mammal, to become helpless in the presence of an enemy, though when family cares are over the conditions might be entirely reversed. Even such timid creatures as rabbits and hares sometimes strenuously defend their young from the attacks of weasels and stoats.

As the fox trotted up the hillside path, the moorhen joined her mate in the tangle of the reeds, and, without fear, wandered [Pg 404] over the marshy ground in the neighbourhood of her nest. Then she swam out across the narrow channel, and settled down, in fancied security, to brood once more over her speckled eggs. She had just taken her accustomed position, when the hedgehog, pushing the reeds aside, became aware of the strong scent on the margin of the pond. The hungry "urchin's" intelligence, though limited, at once suggested that the scent of a mothering bird might lead to a clutch of delicious eggs, or to a brood of plump and juicy nestlings. Following the trail, the hedgehog came to the marshy ground at the margin of the narrow passage where the bird had crossed, and, with head erect, sniffed the tainted wind blowing gently shorewards from the brooding moorhen. In her eagerness, she lifted herself slightly at the edge of the bank, missed her footing, and fell into the pond, not more than two or three feet from the moorhen. The bird, hearing the splash, dived instantly; her mate again came quickly to the scene and tried to lead the enemy away, but the hedgehog, heedless of every artifice, paddled slowly to the platform of dry flags, and helped herself to a repast more appetising than any she had [Pg 405] recently enjoyed, while the birds, flapping their wings, circled angrily about the pond, and pecked vigorously, but vainly, at the marauder's prickly coat.

Late the next evening, the hedgehog discovered a fledgling thrush hidden in the grass beyond the alders. In response to the cry of the young bird, the mother thrush flew straight to the spot, and, with a lucky blow struck full at the hedgehog's snout, so intimidated her enemy that she curled up immediately and allowed the fledgling to escape unharmed.

The tender grass was reaching up to seed, the may blossom was burdening the air with rich perfume, and summer had almost come, when, late one night, the hedgehog, hunting among the shadows of the trees, chanced to hear a low, bleating sound, like the voice of a leveret calling to the mother hare out feeding in the clover. She had never heard that sound before, but its meaning, nevertheless, was plain, and without hesitation she replied. Again the sound broke the stillness, as a dim form lifted itself clumsily from the ditch and came towards her. [Pg 406] Presently she felt an inquiring touch, and, turning, found herself face to face with a male hedgehog that had followed her path through the undergrowth. Nature had not been lavish in his adornment; like the female, he was a plain little creature, brown and grey, fitted to sleep unnoticed among the wind-blown leaves and twigs beside a sheltering mound.

Theirs was an odd and awkward courtship—its language a medley of unmusical squeals and grunts; and if a difference arose it was settled by one curling up into a ball till the other had forgotten the quarrel. But soon they became good friends, hunted together all night and slept together all day, while the year drew on to summer and then, almost imperceptibly, declined. Devoting much of their attention to domestic affairs, they built a large, dry nest among the foxgloves near the stream; where, towards the end of hay harvest, three naked little "urchins" came into the world, to be reared, just as the mother hedgehog herself had been reared, till autumn merged into winter, and winter's cold induced each to go in loneliness and build a snuggery for sleep.

NIGHT IN THE WOODS.

[Pg 407]

I.

[Pg 409]

Top

HAUNTS OF THE BADGER AND THE FOX.

Comparatively little seems to be known of the night side of wild life in this country. Night watching involves prolonged exposure, unremitting vigilance, absolute quietness; and yet, to the most alert observer, it often results in nothing but disappointment and vexation.

Some time ago, during the moonlit nights of several months, I kept watch, near a "set" inhabited by half-a-dozen badgers, a vixen and her cubs, a rabbit and her numerous progeny, and a solitary little buck wood-mouse, whose close acquaintanceship I made after I had captured him in a butterfly-net placed as a spring-trap above his narrow run-way in the grass. This "set"—which I have already partly described, in writing of [Pg 410] Brock, the badger—seemed to be the common lodging house of the wood. Its numerous inhabitants, though not on terms of friendship, were, apparently, not at enmity. The wood-mouse and the rabbits, while entering or leaving the underground passages, and wandering through the paths in the wood, took care to avoid their powerful neighbours; the foxes, believing that out of sight is out of mind, avoided with equal care all chances of encountering the badgers; and the badgers, sluggish in movement and tolerant in disposition, refrained from evicting the foxes or digging out the rabbits.

In the undergrowth, but away from the well worn tracks used by the creatures as they stole out to feed, I had chosen three hiding places, representing in their relative positions the corners of a triangle the centre of which was the main entrance to the "set." I was thus able, whatever might be the direction of the wind, to lie to leeward and obtain a clear view of the principal opening, while I incurred but slight risk of detection, unless the rabbits or the wood-mouse crept into the brambles.

It was during the last week of watching [Pg 411] that my patience received its best rewards. Almost regularly then, as the shadows deepened before moonrise, the rabbits stole out, and, sometimes with no hesitation, sometimes after much cautious reconnoitring and sniffing the air and "drumming" alarm signals on the mound before their door, hopped along the paths towards the clover-fields outside the wood. Soon after the rabbits appeared, the wood-mouse timidly peeped around the corner of the entrance, and, seeing nothing of his enemy, the brown owl, disappeared, with a rustle, among the dead leaves that filled a hollow where the old, disused workings of the "set" had "shrunk."

On several occasions, the vixen led forth her cubs long before the badgers came in view, and while the light yet lingered on the crests of the neighbouring hills. The little family went away silently to a dense furze-brake about a hundred yards distant on the lower edge of the wood, and, till the sun had gone down, remained close-hidden in a lair that I afterwards discovered amid the long grass in the heart of the thicket. [Pg 412]

More frequently, however, I saw nothing of the vixen till nightfall, though the cubs, impatient of confinement, now and again visited the mound outside the "set," and for a few moments played together on the bare soil thrown up by the hard-working badgers, as, in spring, they enlarged their breeding chamber. But, in the first calm hour of night, when the red afterglow had faded from the hills, and the moon, ascending cloudless in the southern sky, cast long, mysterious shadows down the aisles of the wood, the fox-cubs and their dam came boldly out, and, instead of moving off towards the furze, adjourned to a rill close by, whence, after quenching their thirst, they repaired to a glade above the "set," and in this favourite playground frisked and romped, unremittingly guarded from danger by their devoted mother. My presence unsuspected, I watched them, little dim figures, flitting to and fro.

When they had gone far up the winding pathway to the cornfields, and the silence was no longer broken by their low cries of dissembled rage and fear, I sometimes lingered in my hiding place; and as on the [Pg 413] grass I lay, looking towards the stars that twinkled between the motionless leaves of the trees above me, my thoughts went back to a time long before our village had been built beside the river; before Giraldus Cambrensis had journeyed hence with the pilgrim band towards Sant Dewi's shrine; before the great Crag of Vortigern, across the near dingle, had resounded with the blare of the trumpets of war; before even, in the primitive hut-circle on the opposite hill, wild little children had played about the twilight fires kindled in readiness for the home-coming of the weary hunters—a time when the fox, the badger, and the tiny mouse had nightly journeyed through the woods, and the call of the gaunt wolf to his mate had weirdly echoed and re-echoed in the valley, startling the innocent hare in the open waste above the slope, and the busy beaver on the dam below in the pool at the bend of the river.

The badgers—or "earth-pigs" as the country folk have named them—were the original occupants of the "set," unless, however, the earliest excavations had been made by the ancestors of the old doe-rabbit [Pg 414] now inhabiting a side apartment. The foxes and the wood-mouse might have been looked upon as interlopers, but they often played the part of scouts and sentinels, quick to give alarm to the tolerant, easy-going badgers, in case of imminent danger from the visit of a dog or a man to the neighbourhood of their retreat.

The badgers were more irregular as to the time when they left the "set" than were any of the other inhabitants. Perhaps they suspected a human presence, because of some peculiar vibration in the earth through a false step of mine. Perhaps, during certain conditions of the atmosphere, a taint—borne from me, on a wave rather than a current of air, to the wide archway beneath the tree-roots in front of the main entrance, and then drawn down into the draughty passages—was detected by them immediately they passed beyond the stagnant atmosphere of the blind-alley where they slept. Evening after evening, one of the old badgers would appear at the mouth of the "set," and, with snout uplifted in the archway of the tree-roots, would stay as motionless, but for the restless twitching [Pg 415] of the alert nostrils, as were the trees and the stones around his home, while I, not even daring to flick an irritating gnat from my forehead or neck, would wait and long for the philosopher in grey to make up his slow-moving mind.

With regard to the badger's habit of staying for some time in the doorway of his home, it may be mentioned that years afterwards, when one night I compared my notes with those of a companion who had hidden near the main opening of the "set" while I had watched by a hole higher in the wood, I found that each entrance had, simultaneously and for long, been occupied by a vigilant badger; and, as both animals were full-grown "greys," I concluded that parent badgers not unusually took ample precautions against surprise before allowing their cubs to venture out into the night.

Once away from the "set," the old male badger seemed to lose suspicion of any obnoxious presence. Then, lumbering after him, every member of his family would appear in full view on the mound, and, with little fits and starts of pretended rage and fright, would roll over and over each [Pg 416] other, rush helter-skelter back to the underground dwelling and out again, and round and round the tree-trunks. A favourite trick, indulged in by young and old alike, was that of raising themselves on their hind-legs close beside a broad beech-trunk near the "set," and then, on tiptoe, stretching out their fore-claws to the fullest extent and scratching vigorously at the bark.

This trick irresistibly reminded me of an incident connected with a shooting expedition to the moors, when, one evening, after much gossip in the ingle-nook, I accompanied my jolly host to the barn, and there, much to the merriment of all concerned, acted as judge, while, by the light of a lantern, the farmer measured and recorded the height of his wife, as well as of each of his six children and his servants, against the oaken door-post, and finally insisted that he himself, a veritable giant, should submit to the test, and gave orders for a chair to be fetched that "mother," a stout little woman of some sixty inches in height and, also, in circumference, might mount to the level necessary for "chalking his mark." [Pg 417]

One day a keen naturalist and sportsman, whose acquaintance I had recently formed, proposed to join me in my vigil near the badger's home. In the declining afternoon, we left the village, crossed the bridge, and made a detour of the river path. As we passed along, I showed him an otter's "holt" under a shelving bank, where, on the fine, wet sand, the prints of the creature's pads were fresh and clearly outlined. We then visited an "earth" within the wood, in which dwelt a lonely old fox I had often watched as he stole along the rabbit-tracks towards the Crag of Vortigern; and there I pointed out how crafty Reynard, having selected a convenient rabbit burrow, had blocked up every hole—but one, in a thick clump of brambles—with soil thrown out in digging, and how the grass and the ground-ivy had luxuriantly covered the bare mounds, and so encroached on the fox's winding track through the wood and about the bramble clump, that even to an experienced visitor the only fox-sign likely to be detected was in the loose arrangement of the bents and the twigs by the arch of the run-way as it entered the thicket. [Pg 418]

Rabbits, as well as water-voles and field-voles, are particularly careful to nibble off wind-blown or sprouting twigs that encroach on their tracks through the undergrowth; but foxes, otters, and badgers simply brush them aside as they pass.

The sun had not yet gone down when we arrived at the "set." I had planned an early visit, so that my friend might have an opportunity of examining the much frequented track-ways, the footprints of the badgers on the soft earth of the mound, and the scratches on the tree-trunk where the badgers had sharpened their claws and incidentally measured themselves. These numerous claw-marks were especially interesting, and, on a certain tree by the "set," they formed irregular lines extending from a foot above the ground to a height of three feet or rather more. The lowest scratches had been made by the cubs seated on their haunches and facing the tree; a little higher, the marks were those of the parent animals while in a similar position; after a space in which a few abrasions occurred, the marks showed how the cubs had gradually grown till they [Pg 419] could reach within a few inches of the clear, deep furrows scratched by the old male badger as he measured his full length against the tree.

"AS HE MEASURED HIS FULL LENGTH AGAINST THE TREE." (See p. 419). To List

After making observations with the utmost wariness, we hurried away, so that, before dusk, our scent might evaporate, and become almost imperceptible in the vicinity of the principal entrance to the lonely burrow.

After a second ramble by the riverside, we returned in the face of the wind, and at twilight began our silent watch. A robin sang plaintively from the hawthorns on the outskirts of the wood; the rooks sailed slowly above us, and then, gossiping loudly of the day's events, congregated around their nests in the great elms dimly outlined against the pearly southern sky; the wood-pigeons dropped one by one into the beech-trees near us; and a jay, uttering his harsh alarm, hopped in and out of some young hazels fringing the glade beyond the "set." Presently, a brown owl, in a group of tall pines near the little rill that made faint music in the woods, began to mutter and complain, in those low, peculiar notes that are often heard before she leaves her [Pg 420] daytime resting place. Then no sound disturbed the stillness but the far-off cawing of the rooks, and the only creatures visible were some rabbits playing in the moonlit glade, and a glow-worm shining with her soft green light on a bramble spray within my reach.

Nearly half an hour passed by, and no sign of life came from the badgers' home. Then the familiar white and black striped head, framed in the darkness beneath the gnarled tree-root, suddenly appeared, and as suddenly vanished. Another half-hour went by, and yet another, but no further sign was given. My companion, unused to such a long vigil, shifted uneasily, and protested that he was tingling with cramp and longing for sleep; presently, unable to endure his discomfort, he arose, and stretched his limbs before settling down again amid the briars.

Our patience was in vain. Once more the badger came in sight, but my companion did not see what I myself had noticed, for sleep had sealed his tired eyes, and when I nudged him he awoke with such a start that the badger instantly withdrew into the burrow. [Pg 421]

By the glow-worm's lamp, I found from my watch that midnight had long passed; and so, since the hour was towards dawn and the moon was not favourable for close observation of the "earth-pigs," even if they crossed the open glade, I whispered to my friend that the proceedings, in which his interest had manifestly waned, were over for the night. His disappointment was keen, and though to me the night seemed warm, he, accustomed to a tropical climate, chattered with the cold. He had not even noticed the first appearance of the "earth-pig," and henceforth night watching held no charm for him.

My own disappointment, if only for my friend's sake, was also keen; but, on the evening following those hours of fruitless watching, I discovered the vixen's lair in the furze-brake, and learned why she resorted thither with her cubs, before the badger family had awakened from their day-dreams, or the pale glow-worm's rays had lit up the dew-besprinkled spider-webs.

Knowing that badgers are, as the country folk say, *pwdu* (pouty) creatures, likely to sulk at home for several nights if they [Pg 422] consider it unsafe to roam abroad, I carefully examined the mound of earth and the beech-trunk near the "set," that I might learn whether the animals had been out of doors since my previous visit. On the soil, fresh footprints could be seen, their outlines clearly lit and deeply shadowed as the sun sank in the west, and, in some of the scratches on the beech, the pith had barely changed its colour from creamy white to the faintest tinge of brown. I concluded, therefore, that the badgers had been out, as usual, some time before the dawn. My eyes, however, were not sufficiently trained to detect any sure evidence of the recent movements of the vixen and her cubs.

Walking along the tracks, I chanced to notice that the path by which the vixen sought the shelter of the furze-brake branched off at a sharp angle, and led into the thicket at a bend that was hidden from my sight while I watched near the "set." Picking my way in a line straight through the tangle and parallel with this path, I came to an opening where the grass was beaten down for about six square yards—more particularly for two or three yards in the part nearest the spot [Pg 423] at which the tunnelled run-way entered it. Along the margin of this open place, I could find no second entrance; everywhere at the foot of the surrounding gorse-bushes the long grass grew in an unbroken line, except close to the mouth of the run-way. There I found a shallow depression, not unlike the "form" of a hare, but longer and broader, and I determined to keep strict watch evening after evening, till I learned the reason for the occasional visits of the vixen and her cubs to the brake. But I little imagined that the secret would quickly be disclosed, for it was my belief that, should the vixen venture to the mouth of the "set" before the gloom was deepening into night, she would cross the line of my scent, and either move away from the direction of the furze-brake or return to her underground chamber. And yet previous experiences led me to hope that, if certain atmospherical conditions should prevail, the scent would probably become so weak that she would recognise no cause for alarm.

It was the work of a few minutes for me to make couch of grass and twigs behind a screen of broken furze-branches well in [Pg 424] from the grassy opening. Then, by raising with a prong-shaped stake the grass I had trodden down, and by thrusting back the bramble-trails and fern-fronds I had brushed aside, I carefully removed as far as possible all traces of my visit.

I had scarcely settled down to watch and listen, when the faint snap of a twig reached my ears, and I saw that the vixen with her cubs had arrived on the scene. She walked around the enclosure, sniffing now and again in the grass, while the young foxes frisked and gambolled with each other, or trotted demurely by her side. She was at first suspicious, but for some reason she soon gained confidence; then she squatted in her lair, and surrendered herself, with patient motherhood, to be the plaything of her healthy, headstrong youngsters.

For more than a half hour I watched the happy family, the little ones climbing over the mother's back, and licking or biting her ears, her pads, her brush, or racing over the grassy plot, frolicking with each other till some little temper was aroused and play degenerated into a fight. In general, they behaved like wild children [Pg 425] without a thought of care, yet they never went beyond the grass-fringe into the thicket, and to each low note of warning or encouragement from their dam they gave immediate attention. Sometimes the vixen bounded gaily about the edge of the gorse, stooping again and again to snap with pretended rage at one or another of her offspring. But for most of the time she remained in her lair, listening intently for the slightest sound of danger, and guarding the only approach through the bushes.

I longed to discover what she would have done had I suddenly come upon her and cut off her retreat, but I dared not move for fear of raising alarm. It is more than likely that, finding me in the path, she, snarling and hissing, would have dashed without hesitation into any part of the furze-brake, and her young would have followed with desperate haste and vanished at her heels within the shadows.

By-and-by she led her little ones back through the run-way, and when, a few minutes afterwards, I stole to the outer edge of the thicket, I saw the merry family stooping in a row beside the rill, and lapping the [Pg 426] cool, delicious water, which refreshed them after their rough-and-tumble sport. From the rill they wandered off into the gloom beneath the beech-trees, and I, satisfied with having added to my knowledge of the life of the woods, returned homewards in the light of the rising moon.

II.

[Pg 427] Top

THE CRAG OF VORTIGERN.

One of the chief difficulties with which the naturalist has to contend while watching at night is the frequent invisibility of wild creatures among the shadows, even when the full moon is high and unclouded. The contrasts of light and shade are far more marked by night than by day; by night everything seems severely white where the moonbeams glance between the trees, or over the fields, or on the river, and the shadows are colourless, mysterious, profound; whereas by day variety of tone and colour may be observed in both light and shade, and every hour new and unexpected charms are unfolded in bewildering succession.

The wild creatures of the night often seem to be aware of their invisibility in the [Pg 428] gloom, and of the risk they run while crossing open spaces towards trees and hedgerows where an enemy may lurk awaiting their approach. A fox is so familiar with his immediate surroundings that, till his keen senses detect signs of danger, he will roam unconcernedly hither and thither in the dark woods near his "earth," frolicking with his mate, or hunting the rabbits and the mice, or sportively chasing the wind-blown leaves, as if a hound could never disturb his peace. The fox knows the shape of each tree and bush, and of each shadow thrown on the grass; he notes the havoc of the tempest and the work of the forester. When the wind roars loudly in the branches overhead, or the raindrops patter ceaselessly on the dead herbage underfoot, or the mists blot out the vistas of the woods, he seldom wanders far from home, for at such times Nature plays curious tricks with sound and scent and sight, and danger steals upon him unawares.

The hunted creatures of the night so dislike the rain, that during a storm Reynard would have difficulty in obtaining sufficient [Pg 429] food; but down in the river-pools below the wood, fearless Lutra, unaffected by the inclement weather, swims with her cubs from bank to bank, and learns that frogs and fish are as numerous in the time of tempest as when the moon is bright and the air is warm and still.

Since my earliest years of friendship with Ianto the fisherman and Philip the poacher, I have regarded night watching in the woods or by the riverside as a fascinating sport, in which my knowledge of Nature is put to its severest test. By close, patient observation alone, can the naturalist learn the habits of the creatures of the night; and if it should be his good fortune to become the friend of such men as I have mentioned he would find their help of inestimable value.

To Ianto and Philip I owe a debt of gratitude, of which I become increasingly conscious with the passing of the years. I could never make them an adequate return for their kindness; but I am solaced by my recollection that I was able to comfort such staunch old friends when they were passing into the darkness of death—haply to find, beyond, some fair dawn brighter than any [Pg 430] we had together seen from the hills around my home. Often, as I write, I see them sitting in the evening sunlight of my little room; often, in my garden, I see them walking up the path attended by my dogs that now are dead; often, in the river valley, whether I wander by night or by day, I see them at my side.

Ianto and Philip were always eager to help me by every means in their power, but Philip, because of the risk to my health, would never invite me to accompany him when the night was cold and stormy. One afternoon, as Ianto and I were returning home from the riverside, the old fisherman remarked: "I met Philip last night, sir, and he wants you and me to come along with him for a ramble to the woods above the Crag. He's got something to show you; I think it's an old earth-pig that lives in the rocks. What do you say to joining me by the church as soon as you've had something to eat? Then we'll go together as far as the bridge, but I'll leave you there, for I've got a little job on hand that'll keep me till sundown, I think. You'll find Philip at the 'castell' (prehistoric earth-work) above the Crag, and I'll wade the river and be with [Pg 431] you again sometime 'between the lights.' Keep to cover, or to the hedges and the lanes, and look about you well, most of all afore you cross a gap, and when you're going out of cover or into it. Nobody must have a chance of following you to-night to the Crag; so, if you meet a farm labourer sudden-like, make off to the furze by the river farm, and double back through the woods. You'll get to Philip early enough. He's going to net the river after we leave him. It's a game I don't care much for—maybe because I've given it up myself—but I've promised to do something aforehand, that, if Philip didn't want you particular, he'd be bound to do hisself. That's why I'm to leave you at the bridge."

I was tired after a day's hard fishing, but I readily fell in with the arrangements my two old friends had made. On the way to the bridge, Ianto gave me further instructions. "If, when you're nigh the Crag, sir, you happen to come across a farm servant, or even if you think, from seeing a *corgi* (sheep-dog), that a farm servant is near, get right away, and, as soon as you're sure nobody knows where you are, give that [Pg 432] signal I taught you—four quick barks of a terrier with a howl at the end of 'em. Philip'll understand. But if everything goes well till you get to the Crag, make that other signal—the noise of young wood-owls waking up for the night—and Philip's sure to answer with a hoot. Then let him come up to you; but, mind, don't you go to him."

A little mystified by Ianto's last injunction, I crossed the bridge, passed through a succession of grassy lanes that for years had fallen into disuse, picked my footsteps cautiously through the woods, and arrived without adventure at the top of the Crag.

Getting down into the oak-scrub, I stood within the deep shadows at the base of the great rock, and gave the signal—a harsh, unmusical cry, such as a hungry young owl would utter at that time of the evening.

The cry had scarcely gone forth, when I was startled by a voice from some hollow quite close to my side: "I'm Philip. Don't move—don't speak. A man's watching you from the blackthorns at the top of the wood. He hasn't seen me. Don't look his [Pg 433] way, but walk along the path below, and when you reach the end of the wood turn up and hide in the cross-hedges, so that you can watch him if he comes out anywhere in the open. And, mind, don't let him see you then. If he goes back to the farm, give the signal again; or, if I give two hoots, one about ten seconds after the other, come to me, but don't pass this place. The fellow isn't of much account, but we must get rid of him before I can stir. He's kept me here for the last half-hour."

Philip ceased speaking, and I walked carelessly down the wood, pausing here and there to peep through a patch of undergrowth and to satisfy myself that the man at the top of the wood had not moved. When outside the wood, I turned rapidly up the hill and found an excellent hiding place among some brambles on a thick hedge. From this spot I could command a view of the meadows above the wood, and could easily retreat unseen if the farm labourer happened to come towards me.

I watched patiently for twenty minutes or so, then heard Philip's welcome signal from a fir-spinney on the far side of the [Pg 434] Crag, and hastened to his side. In reply to my question as to what had become of the man who had watched from the blackthorn thicket, he pointed to the opposite hillside, where a dim figure could be seen ascending the ploughland in the direction of a distant farmstead. "I expect to be able to show you a badger to-night," he said, "but of course I'm not sure about it. A badger's comings and goings are as uncertain as the weather. But first we'll climb further up the hill. You were asking me about the leaping places of the hares: I know of one of these leaping places, and I think I know of two hares that use them and have lately 'kittled' in snug little 'forms' not far away. We must hurry, else the does will have left the leverets and gone to feed in the clover. You go first. Wait for me in the furze by the pond on the very top of the hill."

When Philip had rejoined me on the hill-top, he rapidly led the way to the fringe of the covert, where he pointed to a low hedge-bank between the gorse and a peat-field partly covered with water. "Hide in the hedge about ten yards from this spot," [Pg 435] he said, "so that you can see on either side of the bank, then watch the path on this side." With a smile he added: "This isn't a bad locality for a fern-owl. So, if you happen to hear the rattle of that bird, you'll know the hare has started from her 'form.'" Then, turning quickly into the furze and taking a bypath through the thickest part of the tangle, Philip left me, and, soon afterwards, I moved to my allotted hiding place.

Before I had waited long, the cry of the fern-owl reached me with astonishing clearness from an adjoining field. Presently, I saw a hare emerge from the gorse and come along the path towards me. At the exact spot indicated by the poacher, she paused, and then with a single bound cleared the wide space between herself and the hedge. With another bound she landed on the marsh beyond, where she splattered away through the shallow water till a dry reed-bed was reached on a slight elevation in the marsh. There she was lost to view; the rank herbage screened her further line of flight.

A minute afterwards, the fern owl's rattle once more broke on the quiet evening, now from a few fields away to my right. For [Pg 436] some time, I closely watched the open space around the hedge-bank, but no animal moved on the path. Suddenly, however, I thought I detected a slight movement in a bracken frond beside the furze. It was not repeated, and I had concluded that it signified nothing, when, to my amazement, I caught sight of a second hare squatting in the middle of the path near the bracken. How she came there I was unable to understand; for some time my eyes had been directed towards the spot, and certainly I had not seen her leave the ferns. She seemed to have risen from the earth—something intangible that had instantly assumed the shape of a living creature. She took a few strides towards my hiding place, but, exactly where the first hare had leaped, she turned sharply at right angles to the path, and with a long, easy bound sprang to the top of the hedge-bank; then with another bound she flung herself into the marshy field. Making straight for the reed-bed, she, too, was soon out of sight.

All that thus happened appeared to be the outcome of long experience; the adoption by the hares of a more perfect plan to mislead [Pg 437] a single enemy pursuing by scent could hardly be conceived. A pack of hounds, "checking" on the path, would in all probability have "cast" around, and, sooner or later, would have struck the line afresh in the marshy field, but a fox or a polecat would surely have been baffled, either at the leaping places or where the hares had crossed through the shallow water.

Man's intelligence, united with the intelligence, the eagerness, the pace, the endurance, and the marvellous powers of scent possessed by a score of hounds, and then pitted against a single creature fleeing for its life, should well nigh inevitably attain its end. Nature has not yet taught her weaklings how to match that powerful combination. And so a naturalist, in studying the artifices adopted by hunted animals, should be interested chiefly as to how such artifices would succeed against pursuers unassisted by human intelligence. I am inclined to believe that even a pack of well-trained harriers would have been unable to follow the doe-hares I have referred to, unless the scent lay unusually well on the surface of the marsh.

I stayed in the covert awhile, but when [Pg 438] the call came for me to rejoin Philip I hastened to the field in which he was waiting. I told him what I had seen, and, together, we paid a visit to the doe-hares' "forms." One of the "forms" lay in a clump of fern and brambles near the corner of a fallow, the other on a slight elevation where a hedger had thrown some "trash" beside a ditch in a field of unripe wheat.

While we stood in the wheat-field, Philip remarked: "We mustn't stay long before going back to the Crag; but I'll call the doe I sent you from this 'form,' and perhaps you'll see one of her tricks to mislead a fox as she returns home. She's very careful of her young till they're about a fortnight old, though soon afterwards she lets them 'fend' for themselves. We'll hide in the ditch, and I'll imitate a leveret's cry. But I mustn't imitate it so that she may think her little one is hurt, else she's as likely as not to come with a rush, and you won't see how she'd act under ordinary circumstances."

When we were comfortably settled in the fern, the poacher twice uttered a feeble, wailing cry, and, after being silent for some [Pg 439] minutes, repeated the quavering call. Then, after a long interval, he again, though in a much lower tone, repeated the cry. No answering cry was heard, but suddenly, as she had appeared on the path by the furze, the doe-hare came in sight at the edge of the ditch a little distance away. She approached for several yards, then disappeared, with two or three long, graceful bounds, into the corn that waved about her as she leaped. She appeared once more, and squatted in the ditch on the other side of the field; hence she jumped high into the air, and alighted on the hedge; then, by a longer bound than any I had previously seen, she gained a spot well out into the field, and raced along, till, directly opposite us, she yet again leaped into the hedge, and from the hedge into the wheat-field, where she immediately lay down with her little ones in the "form."

Ianto, Philip, and I at last settled quietly to watch for the badger's visit to the clearing. Philip told in a whisper of jokes he had played on the keeper; Ianto capped these stories with reminiscences of younger days and nights; and I, though hating bitterly the [Pg 440] ruffian loiterers of the village who subsisted on the spoils of the trap, the snare, and the net, and were guilty of cowardly acts of revenge when checkmated in the very game they chose to play, felt a certain sympathy with the two old men by my side, who, as I was convinced, had fairly and squarely entered into the game, and taken their few reverses without retaliation, only becoming afterwards keener than ever to avoid all interference.

In the height of my enjoyment of an unusually good story, Philip, with a slight movement, drew my attention to a faint, crackling noise coming from the margin of the glade, where moonlight and shadow lay in sharp contrast at the foot of the trees; he then whispered that the old badger was standing there. Ianto almost simultaneously drew my attention thither, but all that I could see at the spot indicated were small, flickering patches of light and shadow.

I quietly drew close to Philip, and murmured in his ear: "Are you sure it's the badger?" He nodded; and I continued, "I see a movement in the leaves, but nothing else." The old man turned his head slightly, [Pg 441] and replied, "What you see is the badger scratching his neck against a tree; the ticks are evidently tickling him." And he chuckled as he recognised his unintentional pun.

For some minutes I could hardly believe he was right; then, slowly, I recognised the shape of the badger's head, and what I had taken to be flickering lights and shadows on the leaves changed to the black and white markings of the creature's face. I had never before seen a badger under similar conditions; and I had often wondered what purpose those boldly contrasted markings could serve. Now, as their purpose was revealed, I was startled by the manifestation of Nature's protective mimicry. Even when, a little later, the animal ventured out from the oak, and stood alert for the least sight or sound or scent of danger, the moonlight and the shadow blended so harmoniously with the white and the black of his face markings, and with the soft blue-grey of his body, that he seemed completely at one with his surroundings, and likely to elude the most observant enemy. Fully a half hour went by before he decided to cross the glade. Then, as if irritated by a sense of his own [Pg 442] timidity, he abandoned his excessive caution, and hastened along his run-way through the clearing; and, as he passed, I noted his queer, rolling gait, and heard his squeaks and grunts as if he were angrily complaining to himself of some recent wrong, and vowing vengeance; I heard, also, the snapping of leaves and twigs beneath his clumsy feet, and I smelt the sure and certain smell of a badger.

Soon, the fisherman and I turned homewards, and left the poacher to less innocent sport. As we gained the crest of the hill, the melancholy cry of the brown owl came to our ears; and Ianto said, "Philip is a big vagabond—bigger than me, I think. No doubt he's fetched his nets from the cave beneath the Crag, and is down at the river by now. Promise me, sir, as you'll never go nigh that cave when he's alive. It's his secret place, as only him and me knows anything about. He told me to ask you that favour."

Long after both Ianto and Philip were dead, I happened one day, while in the woods, to remember the incidents I have just related, and I made my way to the foot of the Crag. I found no opening in the face of the rock, except one—apparently [Pg 443] a rabbit hole— near a rent in the boulder. Climbing around the rock, however, I noticed that a large, flat stone lay in a rather unexpected position on a narrow cleft. I removed it, and saw that it covered the entrance to a dark hollow. At the same moment I heard a slight rustle behind me, as some animal darted from the hole I had previously examined. I scrambled down into the chamber, and there, when my eyes had become accustomed to the darkness, I saw three tiny fox-cubs huddled on the damp, mossy ground. As I knelt to stroke them gently, and my hand rested for a moment on the floor beside them, I touched the remains of an old, rotting net.

FOOTNOTES:

[1] In "Ianto the Fisherman, and Other Sketches of Country Life."

INDEX.

[Pg 445]
Top

- Animals, wild, awakening from hibernation, 146
- — —, — —, dislike rain, 428
- — —, — —, feet made tender by hibernation, 154
- — —, — —, habit of sociable, 160
- — —, — —, keeping to old haunts, 298
- — —, — —, selfishness of, 318
- Ant, habits of queen, 156
- — —, habits of yellow, 65, 66
- Autumn, bird-migration in, 12
-
- Badger, and fox-hounds, 349
- — —, and stoat, 323
- — —, attempt to unearth, 367-373
- — —, fondness of, for honey, 335, 336, 345
- — —, food of, 305, 310-313, 324, 335
- — —, mocked by birds when abroad in daylight, 309
- — —, persecuted for supposed sheep-killing, 353-355
- — —, regular habits in returning to "set" at dawn, 350
- — —, sociability of, 332, 333
- — —, winter habits of, 340, 341
- Badger-cub, and wasps, 337
- — —, caught in trap, 326, 327
- Badger-cubs, at play, 301, 302, 346
- — —, closely confined by parents, 303
- Badger-cubs, dying from distemper, 338
- — —, less nervous than fox-cubs, 321
- Badgers, at play, 359
- — —, carrying bedding to "set," 361
- — —, reconnoitring before young leave "set," 415

- — —, sulking at home if suspicious of danger, 422
- — —, two families inhabiting same "set," 359
- Bank-voles, and kestrel, 147
- — —, colony of, 147
- Basset-hounds, described, 278
- — —, hunting with, 280-282
- Bell, use of, hung round ram's neck, 18
- Blood, significance of fresh-spilt, 75
- Bob, the black-and-tan terrier, 55-62
-

- Character, differences of, in animals of one species, 64
- — —, human, developed by independence of action, 23
- Collie, sheep-killing, 354-356
-

- Dabchick, oar-like wings of, 12
- Ducks, wild, at play, 31
- — —, — —, wedge-shaped flight of, 32
-

- "Earth," fox's artificial, 194
-

- Fear, how it affects wild creatures, 401
- Field-vole, and carrion crow, 165
- — —, and fox, 164
- — —, and kestrel, 148, 149
- — —, and owl, 144, 145, 157, 167, 175
- — —, and weasel, 137, 140
- — —, avoiding rabbit's "creeps," 160
- — —, enemies of, 164
- — —, food of, 137, 142, 143, 154, 155
- — —, hibernation of, 145, 146, 150
- — —, home of, 149
- — —, limbs of, cramped by winter sleep, 153

- — —, restlessness of, in spring, 157
- Field-voles, described, 162
- — —, harvesting seeds, 141, 142
- — —, plague of, 173, 174
- — —, stung to death by adder, 172
- Fox, see also *Vixen*
- — —, and hedgehog, 382-384
- — —, and moorhen, 400
- — —, and wasp, 229
- — —, avoiding traps, 236
- — —, burying rat, 184
- — —, careful not to sleep on straight trail, 237
- — —, careful not to tread on rustling leaves, 220
- — —, entering "breeding-earth" when close pressed, 191
- — —, finding hen's nest in hedgerow, 182
- — —, fight with rival, 227
- — —, hating jays and magpies, 234
- — —, knowledge of the countryside, 238, 428
- — —, luring rabbits, 403
- — —, methods of hunting rabbits, 180
- — —, robbed of spoil by vixen, 183
- — —, seeks mate, 225
- — —, taught by mate, 227
- Fox-cub, chased by lurcher, 222
- — —, cleanly habits of, 212
- — —, described, 203
- — —, food of, 218, 235
- — —, killing hare, 219
- — —, killing polecat, 215, 216
- — —, stealing chickens, 24
- Fox-cubs and partridges, 211
- — —, at play, 412, 422-426
- — —, eagerness of, for flesh, 209
- Foxes, method of preparing "breeding earth," 232
- Fox-hound, "rioting" on cold scent, 189
- Fox-hunt, 186-193

- Frogs, devoured by otters, 35
-

- Geese, wild, 31
- Gipsy, seeking hedgehogs, 387-389
-

- Hare, and renegade cat, 288
- — —, and peregrine falcon, 265, 266
- — —, and poacher, 276, 285, 286
- — —, bravely defends young, 265
- — —, covered with fur at birth, 245
- — —, dislikes entering damp undergrowth, 274
- — —, does not wander far in wet weather, 258
- — —, food of, 248, 249, 251, 260
- — —, "form" described, 245
- — —, killed by lightning, 291
- — —, "leaping places" of, 434
- — —, method of fighting among males, 264
- — —, netted by keeper, 255
- — —, productiveness of, probably influenced by food supply, 276
- — —, recklessness of, in early spring, 263
- — —, running through flock of sheep, 283
- — —, suffers from want of exercise, 259
- — —, suffers less from frost than from rain, 260
- — —, swims across river, 273
- — —, winter habits of, 287
- — —, withholds scent when hard pressed, 283
- Hedgehog, and fox, 382-384
- — —, and moorhens, 400, 401, 403-405
- — —, and owl, 385
- — —, and terrier, 388
- — —, food of, 394, 395, 398, 399
- — —, haunt of, 377
- — —, killing snake, 396, 397

- — —, nest of, 379, 389
- History, vicissitudes of, affecting wild animals, 329
- Hounds, miscellaneous pack, 54, 83
- Hunt, rival, 60
- — —, village, 77, 78, 83
- Huntsman, feeding fox-cubs, 209

- Ianto, the fisherman, 28, 30, 83, 429-442

- Joker, the bob-tailed sheep-dog, 54, 55, 58-60

- Kestrel, attacking field-voles, 148
- — —, preying on bank-voles, 147

- Man, dreaded by wild animals, 13, 40
- — —, senses dulled by immunity from fear, 72
- Mange, attacking carnivorous animals, 212
- March, great changes to wild life in, 263
- Minnows, playing about ledges of rock, 103
- Moorhen, eluding terrier, 61
- — —, killed by otter, 32
- Mouse, singing, 82

- Nature, haunted by Fear, 75
- — —, spirit of restlessness in, 156
- Night, described, 3, 85, 86
- — —, spiritual influence of, 85
- — — -watching, difficulties of, 427
- — — - — —, methods of, 410

- Otter, and big trout, 106
- — —, and dabchick, 12
- — —, and "red" fish, 101
- — —, and water-vole, 86-89, 101
- — —, fighting terrier, 42
- — —, food of, 15, 35, 47, 48
- — —, hunting methods of, 20
- — —, inhabiting drain-pipe, 9
- — —, in winter, 15, 47
- — —, migrating to sea, 46
- — —, playing in heavy stream, 33, 34
- — —, position of, when sleeping, 33
- — —, related to weasel, 22
- — — -cub, capturing salmon, 22
- — —, described, 21
- — —, learns to swim, 9
- — — -cubs, at play, 11
- — — -hounds, 36
- — — -hunt, 37-39, 41-44, 84
- Owl, brown, described, 385, 386
- — —, and fox-cub, 205, 214
- — —, and water-vole, 88, 89
- — —, attacks hedgehog, 385, 386
- — —, preying on field-voles, 157
- Owls, as friends of farmer, 169
- Owls, inhabiting farm buildings, 7
-

- Philip, the poacher, 429-443
- Polecats, enemies of young hedgehogs, 384
-

- Rabbit, burrowing in badgers' "set," 314, 315
- Rabbits, clearing tracks, 418
- Rat, brown, attacked by water-voles, 123
- — —, — —, habits of, 64, 110

- — — -hunting, by riverside, 58-60
- Rats, migration of, 110
- "Redd" of salmon, 99
-

- Salmon, migration of, 95, 96
- — — -fishing, experiences in, 26-30
- — — -pool, seldom visited, 25
- — — -spawn, destroyers of, 99
- — — - — —, guarded by salmon, 98, 100, 101
- Sheep-dog, and otter, 17
- Sorrel, as medicinal herb for wild animals, 335
- Sport, winter, 54
- Squirrel, harvesting only ripe seeds and nuts, 105
- — —, inquisitive, 92
- Stoats, following rats in migration, 110
- Stone-fly, 20
-

- Teal, 31
- Terrier, worsted by otter, 44
- Thrush, autumn song of, 24
- — —, defending young against hedgehog, 405
- Trick, poacher's, to capture hare, 276
- Trout, an old, carnivorous, 95
- — —, habit of, in spring, 19
-

- Viper, attacked by hedgehog, 397
- — —, enemy of young hedgehogs, 398
- Vixen, dispossessing another of "breeding earth," 201
- — —, life spared by hounds, 219
- — —, routing terrier from "breeding earth," 191
- Vixen-cubs, quicker to learn than fox-cubs, 210
- Voles, see *Bank-voles, Field-voles, Water-voles*

-
- Water-shrew, described, 93
- — —, food of, 93, 94, 106, 107
- — —, habits of, 93, 94
- Water-vole, and otter, 86-89
- — —, and owl, 89, 118
- — —, and trout, 94, 125
- — —, as singer, 79-82, 89
- — —, constructing nest, 121, 122
- — —, described, 121
- — —, enemies of, 79
- — —, food of, 71, 105, 106
- — —, habits studied, 80
- — —, home of, 68, 69, 103, 109, 110, 119, 126
- — —, love episodes of, 117-120
- — —, methods of fighting, 119, 123
- — —, winter storehouse of, 105, 109, 126
- Water-voles, attacking brown rat, 123
- Weasel, ferocity of, 76
- — —, food for fox-cub, 213
- Weasels, following rats in migration, 110

PRINTED AT THE EDINBURGH PRESS, 9 AND 11 YOUNG STREET